Marcos Airton de Sousa Freitas

(Editor)

RESILIENT WATER RESOURCES MANAGEMENT:
TOOLS TO DEAL WITH CLIMATE CHANGE AND SUSTAINABLE DEVELOPMENT IN BRAZIL

March 2021 (1st edition)

Cover photo by Marcos A. de S. Freitas; Cerca creek (Paraná river basin), Jardim Botânico, Brasília - DF.

ACKNOWLEDGMENT

To the National Water and Basic Sanitation Agency (ANA), where the author Marcos Airton de S. Freitas has been working as a Senior Water Resources Specialist since 2001. To my parents Benedito da Rocha Freitas Filho and Maria de Jesus Sousa Freitas (*in memoriam*), examples of live and dedication to the family. To my beloved brothers Benedito Neto (*in memoriam*), Marcílio and Márcio.

I would like to dedicate this book to Prof. Dr. Rubem La Laina Porto (*in memoriam*), from the University of São Paulo – USP, and to Eduardo Estellita Cavalcanti Pessôa (Dada), environmental activist, creator of the Center for Advanced Studies of the Cerrado, known as UnB-Cerrado, in Alto Paraíso do Goias (*in memoriam*).

The author and editor Marcos Freitas thanks the other co-authors of this book, namely: Gabriel Belmino Freitas, Márcio Antônio Sousa da Rocha Freitas, Paulo Breno de Moraes Silveira and Sandra Regina Afonso.

Several results presented here were the result of research carried out by the editor and author Marcos A. de S. Freitas, partially with financial support from DAAD (Deutscher Akadmischer Austauschdienst), CAPES (Coordenação de Aperfeiçoamento de Pessoal de Nível Superior) and the Bramar Project - Water Scarcity Mitigation in Northeast Brazil (German-Brazilian research and technology project sponsored by the Federal Ministry of Education and Research - FINEP: Ministério de Ciência, Tecnologia e Inovação).

PRESENTATION

"Before regulatory frameworks, we have to establish civilizational frameworks"
Antonio Prata.

This book entitled **Resilient Water Resources Management** attempts to address the main aspects of integrated management of water resources, climate change and sustainable development in Brazil in recent decades. The book consists of three parts, namely: i) *Integrated Resilient Water Resources Management*; ii) *Climate Resilience in Water Resources* and iii) *Sustainability in Water Resources*.

It is a fact that we are living in an era of great uncertainties related to the climate and that we need tools for climate-resilient water management approaches. In recent decades, climate change is accelerating and disrupting the economies of many countries on a global scale and affecting livelihoods, especially through the impact on water and water-related risks. In this sense, it is imperative to incorporate Climate Risks in Decision Making in the integrated resilient water resources management.

Thus, the first part of the book (chapter 1 and 2) presents a brief analysis of the current model of integrated water resources management in Brazil, with its advances and implementation problems, in order to present proposals for its improvement. The current water resources management model is discussed with an emphasis on aspects of regulation and social control (seeking efficiency and improving quality and rescuing the public sphere as an instrument for exercising citizenship, among other aspects). We sought to analyze the theme from a theoretical-historical approach to regulation (water management models implemented in Brazil and the genesis of regulatory agencies) and participatory management of water resources, involving the relationship of the State and the public sphere, in especially within river basin committees and water agencies.

Next, in chapter 3, a Resilient Approach to Drought Risk Management in the Semi-Arid Northeast of Brazil is presented. This chapter presents the three components of the so-called SIGES (Drought Management System), the items related to drought prediction and monitoring, as well as many reservoir operation methodologies for water scarcity situations. Statistical models and artificial neural networks were used for drought prediction. In order to perform precipitation monitoring, several indexes were adapted and incorporated into a droughts basic characteristic monitoring system (duration, severity and intensity), so that different mitigating actions could be implemented in accordance with the values reached by these parameters. We utilized the following meteorological indexes for this purpose: Rainfall Anomaly Index (RAI); Bhalme & Mooley Drought Index (BMDI); Lamb Rainfall Departure Index (LRDI). Finally, some reservoir operation for water scarcity situations methodologies are presented and discussed. The described components were applied to the Northeast of Brazil, especially Piauí, Ceará and Rio Grande do Norte states.

In the chapter 4 is presented applications of the GAR (1) Model with Fragment Method for Hydrological Drought Risk Assessment in Semiarid Regions. A Gamma Autoregressive – GAR (1) model have been tested and applied, for generating annual flows, coupled with the Fragment Method to disaggregate the annual flows to monthly ones. This coupled model was applied to four typical intermittent basins of the NE-Brazil, with drainage area varying from 410 to 5.695 Km². In order to analyze the performance of the model not only the statistical parameters (mean, variance, lag-1 serial correlation, etc.) of the historical and generated series were examined, but also a storage analysis by mean of the Sequent-Peak-Algorithm (SPA) was performed and additionally the preservation of the droughts and floods characteristics (duration, severity and magnitude) of the historical series was analyzed.

Chapter 5 deals with the Applicability of Multi-seasonal Streamflow Generation Models for Intermittent Rivers. Monthly and annually streamflow generation models have been used for Monte Carlo simulations for design, optimization and risk analysis of multipurpose reservoirs systems.

Chapter 6 addresses the Evaluation of Hydrological Droughts, using the ARRF Model for Monthly Flow Generation in Intermittent Rivers in Northeast Brazil. And it presents the ARR (Alternating Renewal Reward) model for generating annual flow, together with the Fragment model to disaggregate the annual flow into monthly. This coupled model, called the ARRF model, has been applied to three typical intermittent basins in the State of Ceará - Brazil.

In the second part, the book presents the issues related to climate change in Brazil (chapter 7). And it presents several models and methodologies on how to mitigate the effects of extreme events, especially droughts and floods, currently amplified by climate change. In chapter 8 are showed reservoir operation rules as mitigation and adaptation strategies. The case of Paraíba do Sul river basin system is presented.

In Chapter 9 a Drought Monitoring System for the Northeast of Brazil is presented. It is extremely important to have instrumental aid to decision making for reduce disaster risks related to extreme hydrometeorological events such as droughts. In this sense, several indices have been applied to several States of the Northeast Brazil, so that different mitigating actions could be implemented in accordance with the values achieved by these parameters. The following meteorological indices were used: the RAI (Rainfall Anomaly Index), the BMDI (Bhalme & Mooley Drought Index) and LRDI (Lamb Rainfall Departure Index).

The semiarid region of Northeast Brazil experienced during the years 2012 to 2020 one of its worst periods of severe drought. This was largely due to the conditions of the El Niño phenomenon in the Pacific Ocean and the oceanographic and atmospheric conditions of the Atlantic Ocean. In the chapter 10, it was intended to address the aspects of water management due to this extreme drought, both from the supply and demand side, in a typical hydrographic basin in the Brazilian semiarid, the Piancó-Piranhas-Açu river basin.

Chapter 11 aims to present and discuss concepts related to the Annual Average Flood Line (LMEO) and the delimitation of the Permanent Preservation Areas (APP) that aim to protect water courses, considering the institution of the New Forestry Act, the 12.651 Law from May 25, 2012, also called the New Brazilian Forestry Code. The various concepts presented in a series of legal provisions relating to the delimitation of PPAs, aimed at protecting water courses, do not have enough clarity to enable the implementation of public policies in order to ensure the effective conservation of water resources.

Chapter 12 deals with the Rainfall-Runoff Model CN3S for Hydrological Drought Risk Management in Brazilian River Basins. The CN-3S (Curve Number 3 Step) model, with a wide history of application in the semi-arid region of Northeast Brazil, is based on the relationships developed by the US Conservation Service of the CN (Curve Number) curves and it is composed of six calibration parameters.

The Chapter 13 focuses the problem of flood modelling for the Amazonian hydraulic basins, in special, forecast for the city of Manaus. This chapter makes an analysis of historical floods occurred in Manaus, as well as it presents flood forecast models for this city. Linear and not linear regressions models, as well as artificial neural network model for flood forecast have been presented. With those models it is possible to prognose a flood in Manaus station with one month of antecedence.

The third part of this book, called Sustainable Development, brings in chapter 14 the Water-Climate-Forest Nexus and analysis of the Integration of Policies, Governance and Social Participation in Brazil. It deals with the discussion of the integration and implementation of three important Public Policies in Brazil, namely: National Water Resources Policy, National Policy on Climate Change and Forest Policy. The National Water Resources Policy, created by Law 9.433 / 1997, has been implemented throughout Brazil, through its instruments, such as the grant, collection, basin plans, etc. The most recent National Policy on Climate Change has numerous instruments, such as the National Plan and the National Fund on Climate Change.

Additionally, in chapter 15, some aspects related to sustainable development are analyzed, in special, aspects related to the forest bioeconomy and some considerations of forest restoration in river basins. The chapter is entitled Forest Bioeconomy as an Engine for Sustainable Development and Mitigation of the Effects of Climate Change. Brazil presents a great opportunity for the development of the bioeconomy, starting from the management of natural forests, especially public forests, as well as from the integration of the forest component to agricultural systems, especially in private areas. With regards to the management of natural forests, the importance of expanding the use of biodiversity products, especially non-timber products in a sustainable manner and with technological innovation, is highlighted. Currently, only 10 products account for more than 90% of non-timber forest production from native forests. A potential that is still underutilized, especially if we consider the Amazon biome. Regarding the integration of the forestry component into agricultural systems, several forms of production that are being developed all over the world stand out, highlighted here in this chapter, opportunities that are even more interesting when it comes to the Cerrado biome. Finally, these development opportunities from the Forest Bioeconomy stand out as paths to Sustainable Development and the Mitigation of the Effects of Climate Change.

Chapter 16 addresses issues related to the ecological reforestation of ecosystems, especially the cerrado (Brazilian savanna). An important point is the tipping of integumentary dormancy of seeds for forest restoration in the Brazilian cerrado. The Brazilian savanna is among the biomes with the highest floristic diversity on the planet. However, 48% of the native area was removed or altered due to land occupation for agriculture and urbanization. Therefore, there is demand for recovery of the biome. The jatobá (*Hymenaea courbaril*), the mutamba (*Guazuma ulmifolia*) and the canzileiro (*Platypodium elegans*) are species used in the restoration, through the planting of seedlings. However, the seeds of these species show integumentary dormancy necessitating the application of treatments to break dormancy. Different treatments were evaluated aiming at the acceleration, the uniformization of the germination and the increase of the germination rates.

Chapter 17 aims to present and discuss the National Water Resources Management System - SINGREH provided for by Item XIX, of Article 21, of the Federal Constitution of 1988, was created by Law No. 9,433, of January 9, 1997. Among its main objectives, the following can be mentioned: coordinate management integrated water; administratively arbitrate conflicts related to water resources; and implement the National Water Resources Policy. This chapter presents a discussion on the economic and financial sustainability of the National Water Resources Management System - SINGREH, as well as suggestions for its improvement.

Chapter 18 addresses the Application of Multicriteria Analysis in the Study of Operating Alternatives for the Bocaina Reservoir (State of Piauí) and Development of the Region Downstream of the Reservoir (Guaribas River Basin). The Bocaina reservoir, in the semi-arid region of the State of Piauí, with a capacity of 106 hm^3 of water, was built with the objective of regularizing the Guaribas River, among other uses. Over the past three decades, use for irrigation and water supply in neighboring cities has increased significantly. The objective of this study is to analyze, by means of several multicriteria methods, the alternatives of operation of the afore mentioned reservoir for multiple uses. Five alternatives for reservoir operation, socioeconomic and environmental development were analyzed. For this, four different multicriteria techniques were used: i) weighted average method (WAM); ii) compromise programming (CP); iii) Promethee method with weighted averages (Promethee_WAM) and iv) Promethee method.

Therefore, policies related to the management of water resources, the environment and climate change in Brazil are analyzed in a certain way.

The editor.

PART I – INTEGRATED RESILIENT WATER RESOURCES MANAGEMENT

"To the extent that technique and science pervade the institutional spheres of society and thus transform the institutions themselves, the old legitimations crumble"

Jürgen Habermas
Técnica e Ciência como Ideologia.
Lisboa. Edições 70. s.d. p.45

CHAPTER 1

Water Resources Policy in Contemporary Brazil

Marcos Airton de Sousa Freitas
Specialist in Water Resources at the National Water Agency - ANA; Ministry of
Regional Development - MDR

Abstract: This chapter analyzes the water resource management in Brazil, with emphasis on regulation and control (search for efficiency and quality improvement and recovery of the public sphere as an instrument of citizenship). We sought to examine the issue from a theoretical-historical approach (models of government in place in Brazil and genesis of the regulatory agencies) and participatory management, involving the relationship between the state and the sphere public.

Keywords: water resources policy, water resource system, water resources management.

1. Introduction

From the mid-nineties onwards, another form of relationship between governments and society was the possibility for the community to formulate projects and to negotiate directly, or through the mediation of some public agent, the financing for their realization. Such an initiative, however, has also not yielded good results. With rare exceptions, most municipal councils, according to Andrade (2002), is still a fiction, in the sense that these are just pieces in the bureaucratic assembly of government plans and programs without any real support. This is explained by the difficulties existing in our society in establishing spaces for coexistence, discussion, dialogue and representation of different and sometimes divergent interests in situations set up according to principles of equality.

Andrade (2002) concludes that the strong authoritarian and *clientelistic* structure of the regional power system and the absence of a community tradition in our societies appear as strong obstacles to making changes. Where this begins to be overcome, positive signs begin to emerge on the public scene and seeds of true democracy begin to spring up in the arid soil of Northeastern politics.

1.1. Historical Aspects

The National Department of Works Against Drought - DNOCS is the oldest federal institution operating in the Northeast. Created under the name of Inspectorate of Works Against Drought - IOCS through Decree 7.619, of October 21, 1909, edited by then President Nilo Peçanha, it was the first institution to study the semiarid problem. In 1919 (Decree nº 13.687), it was named Federal Inspectorate of Works Against Drought - IFOCS, before assuming its current name, conferred in 1945 (Decree-Law nº 8.846, of 12/28/1945), coming to be transformed into a federal autarchy, through Law No. 4229, of 06/01/1963 (DNOCS, 2009).

This initial institutionalization of Water Administration or Management in Brazil, continued in 1920, with the creation of the Hydraulic Forces Study Commission, of the Geological and Mineralogical Service of the Ministry of Agriculture. In 1933, with the reformulation of this service, the Directorate of Waters was created, which was subsequently transformed into the Water Service. In the following year, that is, in 1934, the Water Service was inserted into the structure of the National Department of Mineral Production, DNPM. In 1940, the Water Service was transformed into a Water Division and the National Department of Works and Sanitation - DNOS was created, which was extinguished in 1990. In 1965, the Water Division was transformed into the National Department of Water and Energy - DNAE, changed name to National Department of Water and Electricity - DNAEE, in 1968.

The increase in demand, coupled with the scarcity and deterioration of the quality of water resources, cause serious conflicts to the multiple use of water, requiring new management paradigms. The Federal Law nº 9.433, of 08.01.1997, which instituted the National Water Resources Policy, has as basic principles, among others, the recognition of water as a vulnerable, finite resource with economic value and the decentralized and participative management of resources water resources. The water management models, that is, the institutional (legal and organizational) and financial mechanisms, have evolved over time in three distinct phases, namely: i) the bureaucratic model; ii) the economic-financial model; and iii) the systemic model of participatory integration. Such models have a close relationship with the models of administration of organizations and with the concept of society of Habermas Theory of Modernity, composed of two "worlds": the "system" (State and market) and the "world of life". Next, the main characteristics of each model will be analyzed, as well as the advantages, disadvantages and problems arising from their implementation in the water resources area (Freitas & Freitas, 1998).

1.1.1. Bureaucratic Model

The planning and management of water resources aim at the prospective assessment of the demands and availability of these resources and their allocation among multiple uses, in order to obtain the maximum economic and social benefits (Barth et al., 1989).

The bureaucratic model had its benchmark in Brazil in the 1930s, with the approval of Decree No. 24,643, of June 10, 1930, better known as the Water Code, with the main characteristics, in administrative terms, of rationalization and hierarchy. At that time, the figure of the public administrator has been predominated complying and enforcing the law.

Due to the need for a large number of regulations, ordinances and rules on the use and protection of water resources, it resulted in authority and power, concentrated in public entities of a bureaucratic nature, presenting the following problems: fragmented view of the water resource management process; centralization of power in the upper echelons, normally distant from the place where the demand for decision was placed; excessive formalism generating control over controls, compliance with rules; difficulties in adapting to changes; standardization in meeting demands; besides attaching little or no importance to the external environment.

As a result of this type of management, public authorities tend to become, in most cases, inefficient and politically fragile in the face of pressure groups interested in concessions, grants, licenses or authorizations. In addition, there is the emergence and aggravation of water use conflicts, resulting in a feedback on the process of drafting laws, asserting the maxim: "if something is not working, it is because there is no appropriate law". On the other hand, the slowness of the justice system (Judiciary Branch), the inoperability and venality of the public branch (Executive Branch), as well as the unscrupulous attitude of some economic agents (entrepreneurs) are blamed for the failure of this model (Lanna, 1997). As Ianni (1963) rightly explains "in a society in which forms and expectations of behavior engendered within patriarchal groups persist and predominate largely, bureaucratization, which implies the adoption of formal behavior expectations, finds no basis for operating effectively".

1.1.2. Economic-Financial Model

The economic-financial model goes back to the economic policy advocated by Keynes, which highlighted the importance of the State's role as an entrepreneurial agent. Such a view was relevant in overcoming the great capitalist depression at the beginning of the century. In the United States, it resulted in the creation of the much-vaunted Tennessee Valley Authority (TVA) in 1933, which greatly influenced the creation in Brazil, in 1948, of the São Francisco and Parnaíba Development Company - CODEVASF, which represents the first Brazilian experience of integrated watershed management.

This model has as main characteristics the use of economic and financial instruments, administered by the public power, in inducing national or regional development, in compliance with the laws in force. The model has its strong point in the implementation of public, sectoral or regional programs, but always within a systemic conception, that is, investment programs in sectors that use water resources are carried out on a sectoral basis: sanitation, irrigation and mainly electrification.

Within the view of Organizational Administration, as highlighted by Lanna (1997), this model has great similarities with the systemic model, since financial and economic instruments are used within a system view. One of the main problems with this model is that of adopting this relatively abstract conception in support of the solution of contingent problems. The application of this model in Brazil resulted in hypertrophied, partial and relatively closed systems, in addition to an asymmetric sectoral development, generating, for example, a great power to use the waters of the main water courses by the electricity sector.

Thus, there is a clear imbalance in the use of water among the various users. As consequences, there are numerous large sectoral development programs (energy program; PLANASA, in the sanitation area; National Irrigation Program - PRONI; POLONORDESTE; Projeto Sertanejo etc.) and creation and strengthening of several regional public entities, such as the DNOCS, SUDENE and others.

According to Lanna (1997), despite the problems resulting from its implementation, this model, even with a visible sectorial focus, can be considered an advance in relation to the bureaucratic model, as it allows the realization of strategic planning of the basin and the channeling financial resources for the implementation of the Hydrographic Basin Master Plans. This rationalization of the economy and the state resulted, therefore, in the hegemony of "instrumental rationality".

Still in the 70s of the last century, the creation of the so-called Hydrographic Basin Committees began. In 1976, the Special Committee was created, the result of an agreement between the Government of the State of São Paulo and the Ministry of Mines and Energy. This committee aimed to promote the solution of the existing problems and conflicts in relation to the uses of water in the Metropolitan Region of São Paulo, as well as the improvement of the sanitary conditions of the waters, an important milestone in the Brazilian water administration, as it promoted interinstitutional and intergovernmental integration for the management of water resources.

As a result of the positive results achieved by the Special Committee, in 1978, through Ordinance No. 90, of March 29, 1978, the Special Committee for Integrated Study of Hydrographic Basins - CEEIBH.

This nationwide committee aimed to promote the rational use of water resources in the river basins of the Union's rivers, through the integration of sectoral studies, developed by the various entities, that interfered in the use of water resources. CEEIBH also had the task of classifying the watercourses under the control of the Union. To achieve its objectives, CEEIBH instituted other executive committees in 1979, in some of the main Brazilian rivers: CEEIVAP, on the Paraíba do Sul River; CEEIPEMA, on the Paranapanema River; CEEIG, on the Guaíba River; CEEIRJ, on the Jari River; and CEEIVASF, on the São Francisco River. In the years 1980, 1981 and 1982, the following executive committees were created, respectively: CEERI, on the Iguaçu River; CEEIPAR, on the Paranaíba River; and CEEIJAPI, in the Jaguari-Piracicaba rivers (Cavalcanti et al., 2008).

After 1988, according to Pompeu (2001), the Union implemented some Hydrographic Basin Integration Committees, namely: the Paraíba do Sul River Basin Integration Committee - CEIVAP (replacing CEEIVAP); the Committee for the Integration of the Hydrographic Basin of the Alto Paraguai Pantanal River - CIBHAPP and the Committee for the Integration of the Hydrographic Basin of the Piranhas-Açu River - CIBHPA. Some of these committees became future embryos of the Basin Committees and Basin Agencies, advocated by Law 9.433 / 97, which instituted the National Water Resources Policy. This was, roughly, a first step towards dialogue with the "world of life".

1.1.3. Systemic Model of Participatory Integration

This model is characterized by the creation of a systemic structure, in the form of an institutional management matrix, responsible for the execution of specific management functions and the adoption of the following instruments: i) strategic planning of the hydrographic basin; ii) decision making through multilateral and decentralized deliberations; and iii) establishment of legal and financial instruments. Its legal framework in Brazil is Federal Law 9.433, of January 8, 1997, also known as the Water Law or the Water Resources Law.

This consideration is the product of an end to the belief in the infinite capacity of the environment to withstand the effects of human activities. Such a change gives credence to public policies - understood as the set of guidelines and actions of a government with a view to achieving certain objectives through instruments for controlling economic activity - the expectation of reversing the current situation of degradation of natural resources. The hydrographic basin strategic planning is a continuous and articulated planning process, within a systemic approach, and based on the study of future scenarios (long-term objectives), with the establishment of specific alternative goals for sustainable development (economic growth, equity and environmental sustainability), with the hydrographic basin as the basic management unit. The deadlines, financial means and necessary legal instruments must be related to each goal.

The hydrographic basin was established as a territorial management unit to the detriment of other political-administrative units - such as municipalities, states and regions - since it integrates the cause-effect relationships that occur in the river drainage network, locus of manifestation of several aspects that the environmental management of this resource must account for (Machado, 2003).

As in long-term planning there is no possibility of obtaining reliable forecasts, there is a need to formulate alternative scenarios for the use, control and protection of water, which will serve as the basis for sectoral plans. In general, it is not possible to establish the most likely scenario, so planning must necessarily be a dynamic process of judgments and decision-making to address new situations in an uncertain future.

The management of water resources is a complex task and involves several conflicting interests. Thus, the public power, without giving up its role as a managing and coordinating body, recognizes the need to promote a decentralization of management, allowing the intervention of representatives of the various segments involved. This is done through social negotiation and the formation of Water User Associations and Hydrographic Basin Committees. The constitution of the Hydrographic Basin Committees aims to promote this social negotiation through the formation of a forum, where all interested parties can clearly and unequivocally expose and discuss their interests.

In other words, public power is not omniscient and cannot be omnipresent. The legal instruments for implementing this model are, among others, the following:

1. River Basin Plans aim to support and guide the implementation of the Water Resources Policy and its management, which are prepared by river basin, by State and for the country. The River Basin Plans are long-term plans, with a planning horizon compatible with the period of implementation of their programs and projects, with the minimum content: i) the diagnosis of the current situation of water resources; ii) analysis of alternatives for demographic growth, evolution of productive activities and changes in land use patterns; iii) balance between availability and future demands of water resources, in quantity and quality, with identification of potential conflicts; iv) goals to rationalize use, increase the quantity and improve the quality of available water resources; v) measures to be taken, programs to be developed and projects to be implemented, in order to meet the expected goals; priorities for granting rights to use water resources; vi) guidelines and criteria for charging for the use of water resources; vii) proposals for the creation of areas subject to restricted use, with a view to protecting water resources.

2. Granting the right to use water resources, which is the administrative act whereby the granting public authority allows the recipient to use water resources, for a specified period, under the terms and conditions expressed in the respective act (Figure 02). The grant confers the right to use water resources subject to water availability and the rationing regime. Therefore, the rights of the following uses of water resources are subject to grant by the Public Power: i) derivation or capture of part of the water existing in a body of water for final consumption, including public supply, or production process input; ii) extraction of water from the underground aquifer for final consumption or production process input; iii) discharge into the body of sewage water and other liquid or gaseous waste, treated or not, for the purpose of dilution, transportation or final disposal; iv) other uses that alter the regime, quantity or quality of water in a body of water.

3. Classification of bodies of water in classes according to the predominant uses, which aims to ensure the quality of water compatible with the most demanding uses for which they are destined and to reduce the costs of combating pollution, through permanent preventive actions. One form of classification was that recommended by Resolution No. 20, of June 18, 1986, of the National Environment Council - CONAMA, replaced by Resolution No. 357, of March 17, 2005, which provides for the classification of bodies of water and environmental guidelines for their classification, as well as establishing the conditions and standards for the discharge of effluents, and revised by Resolution No. 397/2008.

4. Charging for the use of water resources, standing out as an economic-financial instrument, aims to: i) recognize water as an economic good and give the user an indication of its real value; ii) obtain financial resources to finance the programs and interventions contemplated in the water resources plans. The charge for the use and abstraction of water, including the discharge of sewage and other liquid or gaseous waste in the water bodies. Charge thus institutes the polluter-pays or user-pays principle. Charge is therefore no longer a tax.
5. The Water Resources Information System is a system for the collection, treatment, storage and retrieval of information about water resources and factors involved in its management. The basic principles for the operation of the Water Resources Information System are: i) decentralization of data and information collection and production; ii) unified coordination of the system; iii) access to data and information guaranteed to the whole society. The compensation to municipalities and the apportionment of costs for works of multiple use, of common or collective interest, were two other instruments provided for in the bill that were, however, vetoed.

In this model, however, the environment in which the organization operates is emphasized, and how its changing and diverse needs act on the dynamics of the organization, as well as the resulting network of relationships as a result of the demands that arise and the responses issued. In this model, the process of social negotiation in water management is clearly valued, through specific mechanisms and instances for that purpose (Lanna, 1997).

Or as Vattimo (1989) says, "living in this multifaceted world means experiencing freedom as a continuous oscillation between belonging and uprooting".

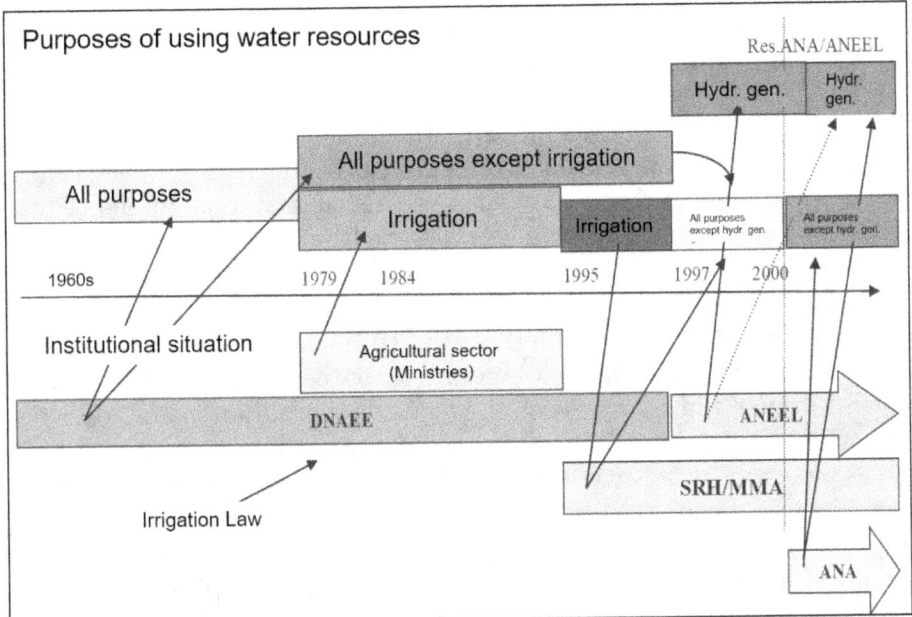

Figure 02: Entities Responsible for Issuing the Right of Use Grant of Water Resources, at the Federal Scope. (Source: www.ana.gov.br)

Kalberg (1980) apud Bin & Castor (2007) identified, in the process of rationalization described in Weberian theory, four types of rationality: i) practice, which suggests an action based on the individual interests of each individual, which develops a capacity to cope with the daily difficulties and learn from them; ii) the theoretical, which involves an understanding of reality through the construction of abstract concepts, instead of the observation of actions, with a characteristic of confronting what is already known with other experiences; iii) the substantive, which appears in the manifestation of man's ability to base his actions on the perception of what is most valuable to him, from not only objective calculations, but also experiences from the past, the present or potential values; and iv) the formal one, which is associated with the calculation and resolution of problems through actions based on rational middle and end standards and on universal rules, laws and regulations.

Habermas (1981) defends a theory that would make it possible to resolve the conflicts in society and, not with a simple solution, but the best solution - one that is the result of the consent of all interested parties. Its greatest relevance is, undoubtedly, in seeking an end to arbitrariness and coercion in the issues that surround the whole community, proposing a way to have a more active and equal participation of all citizens in the disputes that involve them and, concomitantly, obtain the so longed for justice. This form defended by

Habermas is communicative action (theory of communicative action), which branches into communicative action and discourse.

The three water resources management models implemented in Brazil over the past century, were approached, trying to fit them within a historical perspective, showing the different bodies created under their influences, as well as the problems and consequences arising from the implementation of these models in Brazil.

Notably, since the Stockholm Conference in 1972, concerns have increased regarding the use and conservation of natural resources, including water. Since then, several studies and diagnoses on the mode of appropriation and the types of relationships maintained by the actors of the territorial dynamics with natural resources have led to a process of reviewing the attributions of the State, the role of users and the use of water itself (Machado, 2003).

After the 1988 Federal Constitution, the ideas of decentralization and participation took on a new meaning in the Brazilian political-administrative arena, becoming important issues for governments that have sustained the hegemonic point of view on the modernization of the State. These processes of institutional and social changes, introduced through government policies, occur, however, in an extremely varied manner, in view, among other aspects, of the great political-administrative, environmental, social heterogeneity of the different Brazilian hydrographic regions.

Public administration has undergone numerous recent initiatives, in the sense of becoming increasingly transparent, more visible and closer to the population, through the implementation of various instruments of participation and control. According to Torres (2007), a set of factors would partially explain this phenomenon:

- a) the consolidation of Brazilian democracy after 1985, with the improvement of the checks and balances systems between the Executive, Legislative and Judiciary branches;
- b) the active and comprehensive role of the Public Prosecutor's Office, due to the new attributions achieved in CF / 88;
- c) freedom of the press, which enabled the development of more independent and investigative media;
- d) the experiences of participatory budgeting, due to their pedagogical and mobilizing character;
- e) advances in legislation, such as Complementary Law No. 101, of May 4, 2000, also known as the Fiscal Responsibility Law;

f) the gains in transparency brought about by the enormous advances in electronic governance, as is the classic case of the use of the new type of bidding called the auction.

Torres (2007) argues that "in general, due to the deepening of the social division of labor, a process inherent to the development of the bureaucratic model and the capitalist system, public policies become increasingly complex and sophisticated, intelligible only by technicians who developed the necessary expertise to deal with these matters with some security". Another recurring problem is related to the immense volume of information that is generated, within the scope, for example, of water resources plans, technical studies, etc., and made available to civil society. In many cases, civil society finds it difficult to process sophisticated technical information, in addition to the initial difficulty in selecting what is, in fact, relevant and essential to a given problem.

As Torres (2007) evaluates, "a crucial difference must be remembered in this context: raw data and managerial information are totally different things. Usually, a large amount of raw data is made available by the public administration. We find little use in these heaps of information that say nothing to society".

Torres (2007) still states, "generally good information is expensive, requiring high investments, which a good part of the population cannot assume. The expenses necessary to access the internet, to subscribe to newspapers and magazines, cable television, among other available media are high.

Thus, a good part of the population is left without access to information, purely and simply because they cannot pay for it, generating asymmetric and differentiated political behavior among social groups. Naturally, the social sectors more mobilized, organized and with better economic conditions will reach the most valuable information, generating differentiated collective behaviors that privilege the actors with better access to information".

Summarizes Torres (2007): "it is difficult to imagine that a worker who worked hard all day will still dedicate himself to seek some information related to the public administration during his free time, or that he will participate in some meeting in the city council or , yet, to attend a participatory budget assembly in your city (...) ordinary people have little time available and do not want to spend what they have left on political activities that require time, travel, arguments, among other costs required by democracy ".

In the current scenario, fresh water tends to a progressive scarcity, becoming an economic asset itself. This has been occurring due to the pollution of water sources, deforestation, silting of rivers and lakes, the inappropriate use of irrigation, the waterproofing of the soil, the expansion of industrial activities, the irregular occupation of environmental protection areas, the suppression of forests ciliary, inadequate deposition of solid waste, among other numerous actions of modern man.

Given this interrelationship of causes and effects, a broad concept of environmental sanitation was consolidated, which involves water supply, sanitary sewage, urban drainage, solid waste, in addition to vector control (Law nº 11.445, of 05 January 2007).

However, as stated by Machado (2003), the "water policy in Brazil has never privileged sanitation. For more than 60 years, this policy was strongly dominated by the supremacy of power generation, a concern expressed even in the denomination of the national body dedicated to disciplining water use: National Department of Water and Electricity - DNAEE".

Another important legal framework concerns Law No. 12,334, of September 20, 2010, which establishes the *National Policy for Dam Safety* for the accumulation of water for any use, the final or temporary disposal of tailings and the accumulation of industrial waste, creates the Dam Safety National Information System and alters the wording of art. 35 of Law No. 9,433, of January 8, 1997, and of art. 4th of Law 9.984, of July 17, 2000.

After more than two years of debates in the National Congress, Federal Law No. 14,026 / 2020 was published, which provides for new rules for the basic sanitation sector. The new regulatory framework for the sector, approved in the midst of a major health crisis, seeks to modernize and universalize basic sanitation services, which do not reach around 104 million Brazilians, who do not have sewage collection, and 35 million who do not have access to treated water.

Law No. 14,026 / 2020 establishes the new regulatory framework for basic sanitation and promoted several changes in Law No. 11,445 / 2007, which, until then, established national guidelines for basic sanitation. Briefly, Law No. 14,026 of July 15, 2020 updates the legal framework for basic sanitation and amends Law No. 9,984, of July 17, 2000, to give the **National Water and Basic Sanitation Agency (ANA)** the power to edit standards reference on the sanitation service, Law No. 10,768, of November 19, 2003, to change the name and attributions of the position of Specialist in Water Resources, Law No. 11,107, of April 6, 2005, to prohibit the provision by public service program contract referred to in art. 175 of the Federal Constitution, Law No. 11,445, of

January 5, 2007, to improve the structural conditions of basic sanitation in the Country, Law No. 12,305, of August 2, 2010, to deal with the deadlines for the environmentally appropriate final disposition of tailings, Law No. 13,089, of January 12, 2015 (Statute of the Metropolis), to extend its scope to micro-regions, and Law No. 13,529, of December 4, 2017, to authorize the Union to participate in the fund for the exclusive purpose of financing specialized technical services. Some innovations presented by this regulation are listed below.

Concession of basic sanitation services: the provision of public basic sanitation services by an entity that is not part of the administration of the holder depends on the conclusion of a concession contract, through prior bidding, pursuant to art. 175 of the Federal Constitution, its discipline being forbidden by means of a program contract, agreement, term of partnership or other instruments of a precarious nature.

Minimum content of contracts: contracts related to the provision of public basic sanitation services must expressly contain, under penalty of nullity, the essential clauses provided for in art. 23 of Law No. 8,987 / 95, in addition to some provisions, among others: I - goals for expanding services, reducing losses in the distribution of treated water, quality in the provision of services, efficiency and rational use of water, energy and other natural resources, the reuse of sanitary effluents and the use of rainwater, in accordance with the services to be provided; II - possible sources of alternative, complementary or accessory revenue, as well as those from associated projects, including, among others, the disposal and use of sanitary effluents for the production of reused water, with the possibility that the revenues are shared between the contractor and the contractor, if applicable.

Universalization goals: contracts for the provision of public basic sanitation services must define universalization goals that guarantee the supply, up to 12/31/2033, of 99% of the population with drinking water; and 90% of the population with sewage collection and treatment.

Public basic sanitation services will be regulated by the Water and Basic Sanitation National Agency (ANA), which has a legal nature as an autarchy under a special regime, linked to the Ministry of Regional Development and part of the National Water Resources Management System (SINGREH). ANA has administrative, budgetary and financial autonomy; and decision-making independence.

2. Water Resources System Description

The aspect of integrated and collegiate public management deserves special attention and it will be the focus of more detailed analysis here, as it constitutes an instrument of institutional framework for conflicts, inevitable in a continental country with physiographic, hydrographic, geomorphological, hydrological, socioeconomic and large diversity. inequalities and social injustices. It is a concept that was consolidated relatively recently, but goes back to the countless social movements, which, since the 1970s, have been part of the Brazilian political reality.

Thus, the Federal Constitution of 1988 provided, in its Article 21, for the organization of the National Water Resources Management System - SINGREH, formed by a set of legal and administrative mechanisms with the aim of coordinating the integrated management of water resources and the implement, in a participatory manner, the National Water Resources Policy, defined by Law 9,433, of January 8, 1997 (Figure 03).

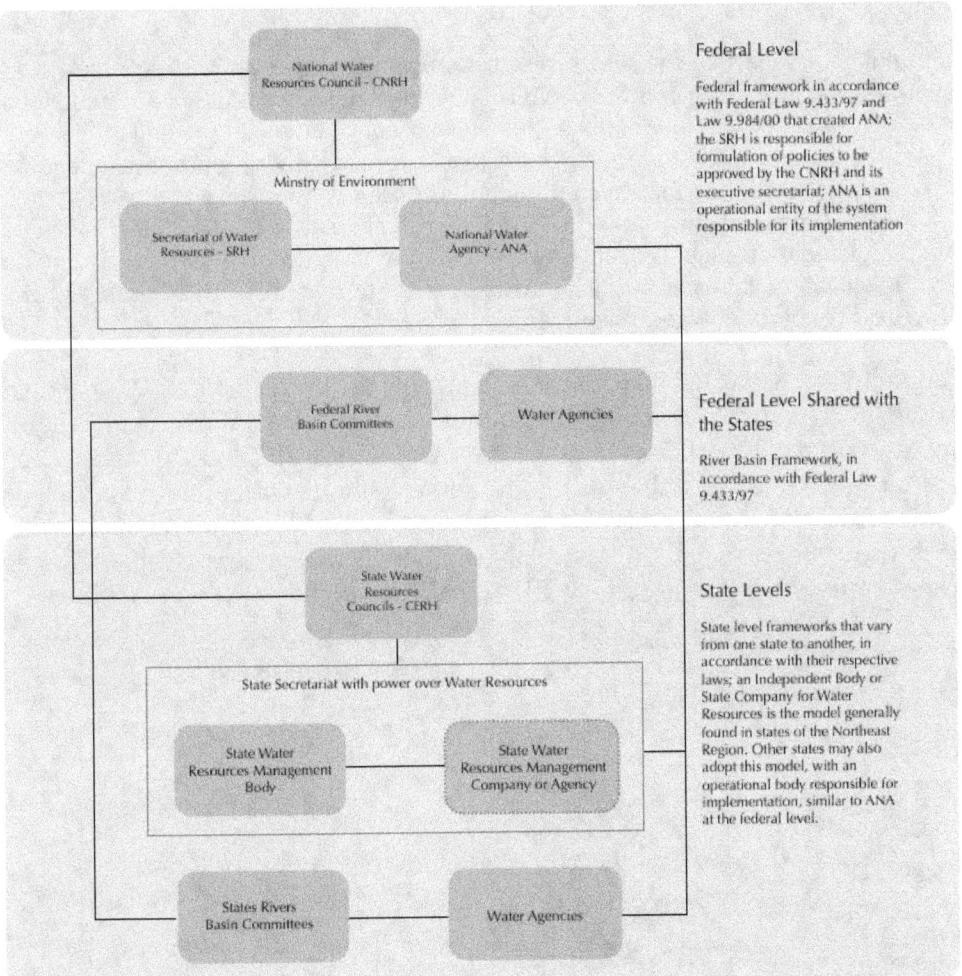

Figure 03: National Water Resources Management System – SINGREH (Source: *GEO Brazil Water Resources*, 2007).

The fundamentals and concepts related to the National Water Resources Management System and water resources management are the result of several conferences held in the last decades. According to *GEO Brazil Water Resources* (2007), "the ten commandments emanating from the Third World Water Forum (Kyoto, 16-23 March 2006) mirror a generic model that today is pursued in many countries, with extreme proximity to the Law National No. 9,433 / 97, without the main practical challenges having been effectively overcome". It follows the study stating that

"The consensus that these declarations assume and the wide dissemination of them by the various multilateral organizations end up making them an obligatory part of the water resources agenda. However, it should be noted that, despite being consensual in relation to the positions of governments of a significant part of the nations, this view is not unique and presents divergent aspects in relation to other perspectives expressed by significant segments of society. More than that, even though the conceptual basis is widely accepted, it must be recognized that no country, developed or developing, has fully implemented such concepts, either due to practical limitations or due to mismatches between its legal structures. state administrative bases and the bases of modern water resources management".

A landmark of the Water Policy in the country is the Water Code (1934), which defined the waters in public, common and private. With the Federal Constitution of 1988, the surface waters (Art. 20, III, VIII and IX) are surface waters (they bathe more than one State, serve as limits with other countries or are the result of Union works), hydraulic potentials and mineral waters; as goods of the States (Art. 26, I) are listed the surface waters (water sources located entirely within the State or the DF) and groundwater (Figure 04).

Water domain

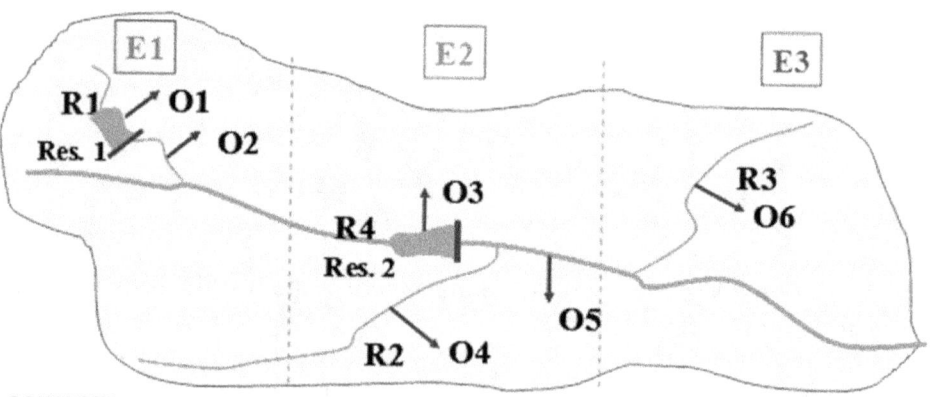

CONCEPTS:
1. The waters of the R4 and R2 rivers and the waters of the Res. 2 reservoir are federal domain;
2. The waters of the R1 and R3 rivers are state-owned;
3. If the Res. 1 reservoir belongs to the Federal Government, the waters will be in the federal domain, otherwise they will be in the state domain.

COMPETENCE:
1. O2 and O6 grants are state;
2. 2. O3, O4 and O5 grants are federal;
3. 3. The O1 grant will be federal if the reservoir is from the Federal Government.

Figure 04: The Water Domain in Brazil (Source: www.ana.gov.br)

This double domination of water, however, created many uncertainties as to the role of the different levels of the federation in the management of water resources. In addition, it is easy to see that most states lack technical and administrative capacity in crucial aspects for the operationalization of committee decisions, particularly, the implementation of systems related to the granting, information, monitoring and inspection (Abers & Jorge, 2005).

2.1 The National Water Agency – ANA

The National Water Agency, in accordance with Article 1, of its Internal Regulation, is an autarchy under a special regime, created by Law No. 9,984, of July 17, 2000, endowed with administrative and financial autonomy, linked to the Ministry of the Environment, is part of the National Water Resources Management System and aims to implement, within its sphere of competence, the National Water Resources Policy, under the terms of Law No. 9,433, of January 8, 1997. After Law No. 14,026, of July 15, 2020, which updates the legal regime for basic sanitation and amends Law No. 9,984, of July 17, 2000, the National Water Agency is renamed the National Water and Basic Sanitation Agency (ANA).

ANA's performance will obey, according to article 1 of its Internal Regulation, the fundamentals, objectives, guidelines and instruments of the National Water Resources Policy and will be developed in conjunction with public and private bodies and entities that are part of the National Water Resources Management System - SINGREH, being responsible for:

I - to supervise, control and evaluate the actions and activities resulting from the fulfillment of the federal legislation pertinent to water resources;
II - to discipline, in a normative character, by means of a resolution of the Collegiate Board, the implementation, operation, control and evaluation of the instruments of the National Water Resources Policy;
III - supervise the implementation of the National Water Resources Plan and participate in studies aimed at its improvement;
IV - provide support for the preparation of water resources plans for basins and hydrographic regions;
V - grant the right to use water resources in bodies of water belonging to the Federal Government, including for the use of hydraulic energy potential;
VI - supervise, with police powers, the use of water resources in bodies of water owned by the Union;
VII - prepare technical studies to support the definition, by the National Water Resources Council, of the amounts to be charged for the use of water resources in the domain of the Union, based on the mechanisms and quantities suggested by the hydrographic basin committees, in the form of art. 38, VI, of Law No. 9,433, of 1997;

VIII - stimulate and support initiatives aimed at creating river basin committees;

IX - implement, in conjunction with the hydrographic basin committees, charging for the use of water resources in the domain of the Union;

X - collect, spend and apply whatever is proper to it and distribute, for application, the revenues earned through the collection for the use of water resources in the domain of the Union, as provided by law;

XI - plan and promote actions aimed at preventing or minimizing the effects of droughts and floods, within the scope of the National Water Resources Management System, in conjunction with the central body of the National Civil Defense System, in support of States and Municipalities;

XII - declare bodies of water under preventive rationing and apply the necessary measures to ensure their priority uses, in accordance with the criteria established in a federal decree, after consulting the respective hydrographic basin committees, if any;

XIII - to promote the elaboration of studies to subsidize the application of Union financial resources in works and services for the regularization of water courses, for the allocation and distribution of water, and for the control of water pollution, in line with what is established in the resource plans waters of the respective hydrographic basins;

XIV - to define and inspect the conditions of operation of reservoirs by public and private agents, aiming to guarantee the multiple use of water resources, as established in the water resource plans of the respective hydrographic basins, in compliance with the provisions of art. 4, § 3, of Law 9.984, of 2000;

XV - to discipline, in a normative character, and authorize the addition of raw water that involves water resources in the domain of the Union, including by setting efficiency standards for the provision of the respective service;

XVI - promote the coordination of activities developed within the scope of the national hydrometeorological network, in conjunction with public and private bodies and entities that are part of or that are users of it;

XVII - organize, implement and manage the National Water Resources Information System;

XVIII - stimulate research, technological development and the training of people for the management of water resources;

XIX - provide support to States in the creation of water resource management bodies;

XX - to propose to the National Water Resources Council the establishment of incentives, including financial ones, for the qualitative and quantitative conservation of water resources;

XXI - promote exchanges with national and international entities related to water resources;

XXII - represent Brazil in international water resources organizations, in conjunction with the Ministry of Foreign Affairs and with other bodies and entities involved;

XXIII - enter into agreements and contracts with federal, state, municipal bodies and entities and with legal entities governed by private law, involving matters related to water resources within its competence; and XXIV - develop and coordinate projects related to the use of water resources supported by national and international organizations. Single paragraph. In carrying out the competence referred to in item II of this article, the respective agreements and treaties will be considered, in the case of river basins shared with other countries.

Table 09, below, shows the number of normative acts issued by the National Water Agency, since its creation.

Table 09: Quantity of normative acts issued by the National Water and Basic Sanitation Agency (ANA)

year	normative acts	trip authorization to leave country	CERTOH	DRDH	water grant	sum
2001	28	17	0	0	102	147
2002	34	52	2	0	318	406
2003	17	46	0	3	389	455
2004	45	54	0	3	602	704
2005	20	22	5	10	314	371
2006	33	42	6	9	504	594
2007	31	40	14	3	511	599
2008	24	55	11	8	751	849
total	232	328	38	36	3491	4125

According to BRASIL (2008), the National Water and Basic Sanitation Agency (ANA) developed the following actions and activities during 2008, namely: Strategic Plan for Water Resources in the Amazon Basin - Right Bank Affluents; Strategic Water Resources Plan for the Tocantins-Araguaia Rivers; Integrated Water Resources Plan for the Doce River Basin (PIRH-Doce); among countless others.

In line with BRAZIL (2008), ANA, in relation to the Strategic Plan for Water Resources of the Amazon Basin - Right Bank Affluents, held, in 2008, "the second stage of overflowing the basins, promoted two rounds of technical meetings with agencies managers and state departments with territory in the basins (AM, AC, RO, PA and MT), integrated data on the affluent basins, elaborated the integrated diagnosis and part of the set of maps that will integrate the study ".

According to BRASIL (2008), the Strategic Plan for Water Resources of the Tocantins-Araguaia Rivers was "elaborated in a participatory way through open public meetings, with the state water resources councils of the states that integrate the region, and by the constitution of a technical monitoring group formed by representatives of the federal and state governments, civil society and water users".

One of ANA's duties is to train organized segments of Brazilian society for effective and efficient participation in the implementation of the National Water Resources Policy. In this sense, in 2008, "about six thousand people were trained, promoting awareness and stimulating the debate on conservation and rational use of water resources" (BRASIL, 2008). In 2006, there were 2,336 people, while in 2007, 2,074 people were trained.

Table 10: Number of Regulatory Experts (January 2007 and June 2009)

Agency	January 1, 2007		June 1, 2009		total expected positions
	number of public servants	% of occupancy	number of public servants	% of occupancy	
ANVISA	451	55.7	661	81.6	810
ANEEL	81	22.2	255	69.9	365
ANS	201	59.1	322	94.7	340
ANATEL	305	42.4	511	71.0	720
ANP	190	39.2	300	61.9	485
ANTT	99	16.8	174	29.5	590
ANA	120	45.1	127	47.7	266
ANTAQ	71	32.3	65	29.5	220
ANCINE	62	41.3	72	48.0	150
ANAC			383	41.5	922
total or average	1580	39.3	2870	57.5	4868

Source: Personnel Statistical Bulletin No. 129, v. 12 and nº 158, v. 14, from the Ministry of Planning, Budget and Management.

In relation to human resources, Table 10 presents the number of specialists in regulation of the various agencies in January 2007. The number of positions in the National Water and Basic Sanitation Agency (ANA) has created by Law No. 10,768 / 2003: i) 239 positions in the Specialist career in Water Resources; ii) 27 positions in Geoprocessing Specialist; iii) 84 positions of Administrative Analyst and iv) 40 positions of administrative technician (medium level), created by Law No. 10,871 / 2004. Of this total, in January 2007, there were only 45.1% of the positions held.

Figure 05 presents a comparison of the situation of occupation of public servants at the National Water Agency, in 2001 and in 2008.

year	temporary employees	permanent employees	commissioned positions	interns (graduates)	outsourced	total
2001	55	3	88	21	167	334
2008	0	169	90	34	236	529

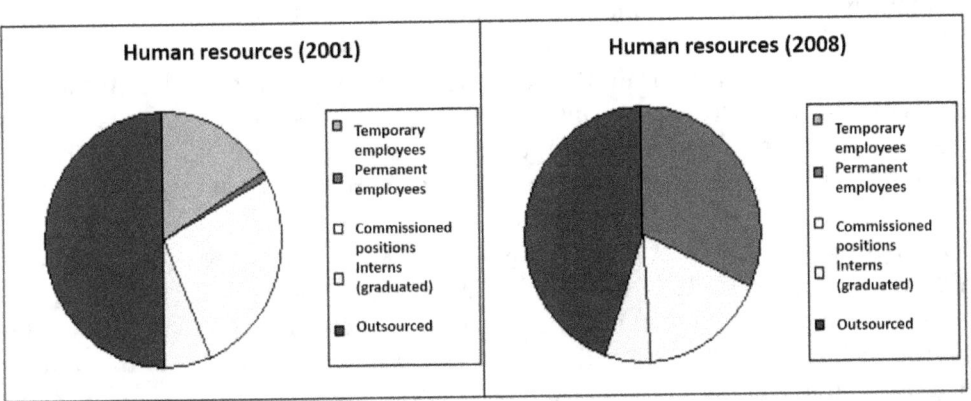

* Among these, 50 have a commissioned position.

Figure 05: Comparison of the situation of occupation of servers at the National Water Agency, in 2001 and in 2008.

It is necessary to pay attention to the following aspects: i) the replacement of temporary contracts with permanent assets (in 36 months, a specific framework should be created - tender or redistribution); ii) the considerable increase in outsourced workers (despite the contests held, especially in the administrative areas; until now, there has been no competition to hire permanent mid-level employees, despite the legal provision of 45 positions) and iii) the occupation of part positions commissioned by permanent assets.

The National Water Agency has the direction of a collegiate body with the characteristics: i) it has 5 members appointed by the President of the Republic; ii) one of them is the top manager of the body: the Chief Executive Officer (also chosen by PR); iii) 4-year terms, not coincident; iv) deliberation by simple majority of votes (requirement: minimum of 3, with 2 of them being the Chief Executive Officer and his successor; v) dismissal without reason only in the first 4 months of the term (after this period, they only lose their term in the event of resignation , final judicial conviction or final decision in Disciplinary Administrative Proceedings - PAD).

Some issues under discussion within the Agencies are: i) the creation of a career development plan and specific salary structure for the staff of regulatory agencies (still under negotiation); ii) wages incompatible with careers of the State and those of the regulated private sectors - the category that demands matching the careers of the management cycle - adverse selection problem and inability to attract qualified personnel; iii) there is no regulation for the Qualification Bonus; iv) the Performance Bonus was regulated with a 25-month delay (the legal forecast was 180 days); v) the *constitutionalization* of the budgetary, administrative and financial autonomy of the Regulatory Agencies is urgent; vi) requested servers discourage the development of independent and autonomous capacities within the regulatory bodies, impairing commitment to the agency's mission; vii) unconstitutionality in the use of civil servants who are already in the exercise of activities in the branches (commissioned, hired on a temporary basis, requested or redistributed from other bodies).

The institutes that impact the Human Resources sector of the Agencies are as follows: i) Law no. 10,871 / 2004: on the creation of careers and the organization of effective positions of special autarchies (ANA); ii) Law no. 10,768 / 2002, on ANA's Staff; iii) Laws no. 9,986 / 2000, on the human resources management of Regulatory Agencies; iv) Law no. 9.984 / 2000: creation of the agency (stipulates a period of 36 months for setting up its own staff); v) Internal Regulations: resolution no. 173/2006 and 348/2007; vi) Organizational Structure: resolution no. 223/2006 and 121/2007. As positive actions of an administrative nature, the following can be mentioned: i) installation of a medical service, with the objective of providing emergency medical care to all employees working at the Agency; ii) the approval, on April 27, 2009, of the Progression and Promotion Regulation, aimed at good functional performance.

The relationship between the supervisory body, MMA - Ministry of the Environment, and ANA - National Water Agency should, according to Presidential Decree No. 3,692 (12/19/2000) and ANA Board Resolutions 173 (04/17/17) / 2006 and nº 348 (08/20/2007), be governed by Management Contract, but so far not implemented. Decree No. 3,692, of December 19, 2000, provides for the installation, approves the Regulatory Structure and the Statement of Commissioned Positions and Technical Commissioned Positions of the National Water Agency - ANA. In its article 10 (Of the Management Contract) it declares "the administration of ANA will be governed by management contract, negotiated between its Chief Executive Officer and the Minister of State for the Environment, within a maximum period of one hundred and twenty days after the appointment by ANA's Chief Executive Officer § 1 The management contract will establish the indicators that allow objectively assessing ANA's performance. § 2 The absence of the Management Contract will not prevent ANA's normal performance in the exercise of its powers".

With regards to the budgetary issue of the National Water Agency, according to Art. 20 of the ANA creation law, ANA's revenues are:

I - the resources transferred to it as a result of appropriations allocated in the General Budget of the Union, special credits, additional credits and transfers and transfers that are granted to it;

II - the resources resulting from the charge for the use of water from water bodies in the domain of the Union, respecting the forms and limits of application provided for in art. 22 of Law No. 9,433, of 1997;

III - funds from agreements, agreements or contracts entered into with national or international entities, bodies or companies;

IV - donations, bequests, subsidies and other resources allocated to it;

V - the proceeds from the sale of publications, technical material, data and information, including for the purpose of public bidding, administrative fees and registration fees for competitions;

VI - remuneration for services of any nature provided to third parties;

VII - the product resulting from the collection of fines applied as a result of inspection actions referred to in arts. 49 and 50 of Law No. 9,433, of 1997;

VIII - the amounts calculated from the sale or rent of movable and immovable property owned by it;

IX - the proceeds from the sale of assets, objects and instruments used for the practice of infractions, as well as the offenders' assets, to be seized as a result of the exercise of police power and incorporated into the autarchy's assets, under the terms of a judicial decision; and

X - the resources resulting from the collection of administrative fees.

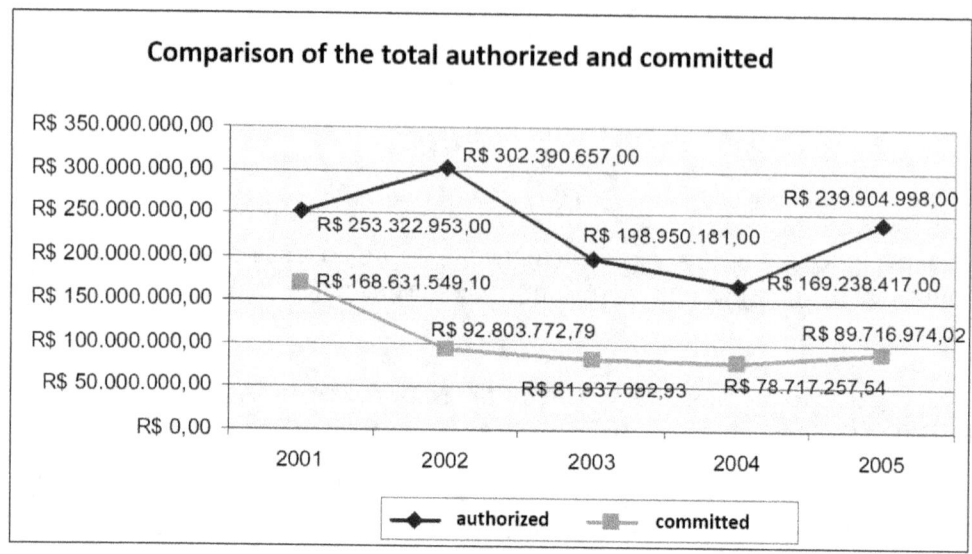

Figure 06: Comparison of the total authorized and committed (National Water and Basic Sanitation Agency - ANA) for the period 2001-2005, in R$ (Source: Siga Brasil).

Article 21 states that the revenues from the collection for the use of water resources in the domain of the Union will be kept at the disposal of ANA, in the National Treasury Single Account, as long as they are not destined for the respective schedules.

§ 1 ANA will maintain records that allow the correlation of revenues with the hydrographic basins in which they were generated, in order to comply with the provisions of art. 22 of Law No. 9,433, of 1997.
§ 2 The cash equivalents referred to in the caput of this article may be maintained in financial investments, in the manner regulated by the Ministry of Finance.
§ 3 (VETOED)
§ 4 The priorities for the application of resources referred to in the caput of art. 22 of Law No. 9,433, of 1997, will be defined by the National Water Resources Council, in conjunction with the respective river basin committees.

Figure 06 shows the comparison of the authorized total and the committed amount (National Water Agency - ANA) for the period 2001-2008. There is a considerable reduction in the budget in the years 2003 and 2004, given the change of government that occurred and the new political orientation in relation to regulatory agencies.

2.2 The Secretariat of Water Resources - SRH / MMA

In 1995, the Water Resources Secretariat - SRH was created, as an integral part of the basic structure of the Ministry of the Environment and the National Water Resources Management System, through Provisional Measure 813, converted into Law 9,649, of 27 May 1998, with the function of formulating the National Water Resources Policy, as well as monitoring and monitoring its implementation. With Decree No. 6,101, of April 26, 2007, SRH changed its name to the Secretariat of Water Resources and Urban Environment, with the following powers:

I - propose the formulation of the National Water Resources Policy, as well as follow and monitor its implementation, under the terms of Law No. 9,433, of January 8, 1997, and Law No. 9,984, of July 17, 2000;

II - propose policies, plans and standards and define strategies on topics related to: a) the integrated management of the multiple sustainable use of water resources; b) the management of transboundary waters; c) the management of water resources in international forums; d) the implementation of the National Water Resources Management System; e) sanitation and revitalization of river basins; f) urban environmental policy; g) urban environmental management; h) the development and improvement of local and regional planning and management instruments that incorporate the environmental variable; i) the assessment and mitigation of environmental vulnerabilities and weaknesses in urban areas; j) pollution control and mitigation in urban areas; and l) integrated management of solid urban waste;

III - monitor the implementation of the National Water Resources Plan;

(...)

VIII - monitor the functioning of the National Water Resources Management System;

(...)

XI - develop actions to support the constitution of the Hydrographic Basin Committees;

(...)

XV - provide technical support to the Minister of State in monitoring compliance with the goals set out in the management contract signed between the Ministry and ANA and other management agreements related to water resources;

XVI - exercise the function of executive secretary of the National Water Resources Council;

(...)

XIX - perform other activities that are assigned to him in the area of his performance.

2.3 The National Water Resources Council – CNRH

In accordance with Art. 1, of Decree No. 4,613, of March 11, 2003, the National Council for Water Resources, a consultative and deliberative body, part of the regulatory structure of the Ministry of the Environment, is responsible for:

I - promote the articulation of water resources planning with national, regional, state and user sector planning;

II - arbitrate, in the last administrative instance, the conflicts existing between State Water Resources Councils;

(...)

VII - to approve proposals for the institution of Hydrographic Basin Committees and to establish general criteria for the elaboration of its regulations;

VIII - to deliberate on the administrative appeals brought against it;

IX - monitor the execution and approve the National Water Resources Plan and determine the measures necessary to meet its goals;

X - establish general criteria for granting the right to use water resources and for charging for their use;

(...)

XII - formulate the National Water Resources Policy under the terms of Law No. 9,433, of January 8, 1997, and of art. 2nd of Law 9.984, of July 17, 2000;

(...)

XIV - define the amounts to be charged for the use of water resources owned by the Federal Government, pursuant to item VI of art. 4th of Law 9.984, of 2000;

XV - to define, in articulation with the Hydrographic Basin Committees, the priorities for the application of the resources referred to in the caput of art. 22 of Law No. 9,433, of 1997, under the terms of § 4 of art. 21 of Law No. 9,984, of 2000;

XVI - authorize the creation of Water Agencies, under the terms of the sole paragraph of art. 42 and art. 43 of Law No. 9,433, of 1997; In accordance with Art. 37, of Decree No. 6.1010, of April 26, 2007, the National Water Resources Council is responsible for exercising the powers established in art. 35 of Law No. 9,433, of 1997.

According to BRASIL (2008), upon the approval of the National Water Resources Plan, the National Water Resources Council instructed ANA to prepare, annually, the Report on the Situation of Water Resources in Brazil, with the objective of making information public respect of the status of water uses, availability, quality and demand. The first edition of this report was launched in the first half of 2009. The complete material is easily accessible and can be consulted on the internet.

2.4 The River Basin Committees

According to ANA Resolution No. 121, of April 23, 2007, in its Article 43, it is incumbent upon the Management of Water Resources Management - GERHI:

> I - propose strategies and mechanisms to support the creation, installation and operation of River Basin Committees and Water Agencies;
> II - articulate strategies for institutional strengthening of SINGREH entities, aiming at the integrated management of water resources in basins and hydrographic regions; and
> III - promote, together with the Hydrographic Basin Committees, the process of negotiating the definition of the Water Agency model and the management contract."

Currently, there are 7 River Basin Committees installed in rivers in the Union's domain, namely:

- Rio Doce Hydrographic Basin Committee (http: //www.riodoce.cbh.gov.br/)
- Paraíba do Sul Basin Committee (http://ceivap.org.br/index1.php);
- Paranaíba River Basin Committee (http://www.paranaiba.cbh.gov.br/);
- Piracicaba Capivari and Jundiaí - PCJ Rivers Basin Committee, (http://www.comitepcj.sp.gov.br/comitespcj.htm);
- São Francisco River Basin Committee (http://www.saofrancisco.cbh.gov.br/);
- Rio Verde Grande River Basin Committee (http://www.verdegrande.cbh.gov.br/);
- Piranhas-Açu River Basin Committee (http://www.piranhasacu.cbh.gov.br/).

2.5 River Basin Agencies

The Water Agencies are executive technical entities that will act in support of the executive secretariat of the basin committees and should contribute all technical subsidies to the discussion on the planning and management of uses in those hydrographic basins (these attributions are provided for in articles 41 and 44 of Law No. 9,433, of January 8, 1997).

The creation of Water Agencies is authorized by the National Water Resources Council or by the State Water Resources Councils upon request from one or more river basin committees. This creation is conditioned, therefore, to the previous existence of the respective committees and to the financial viability ensured by charging for the use of water resources in its area of operation.

Law No. 10,881 / 2004 allows the functions of Water Agencies to be performed by delegating entities. These entities must be included among those provided for in art. 47 of Law 9,433, among civil non-profit organizations and, indicated by the committees, may be qualified by the National Water Resources Council - CNRH for the exercise of legal attributions.

Art. 47. For the purposes of this Law, civil water resources organizations are considered:

I - inter-municipal consortia and hydrographic basin associations;
II - regional, local or sectorial associations of users of water resources;
III - technical and teaching and research organizations with an interest in the area of water resources;
IV - non-governmental organizations with the purpose of defending diffuse and collective interests of society;
V - other organizations recognized by the National Council or by the State Water Resources Councils.

Art. 48. In order to integrate the National Water Resources System, civil water resources organizations must be legally constituted.

By the end of 2008, there were two delegating entities with functions of water agencies installed and in operation - AGEVAP, in the Paraíba do Sul river basin, and PCJ Agency, in the Piracicaba, Capivari and Jundiaí (PCJ) river basins. In the context of Paraíba do Sul, the novelty was the recognition, by the committees of the rivers Pomba, Muriaé, Preto and Paraibuna, of AGEVAP as an entity equivalent to the water agency (BRASIL, 2008).

2.5.1 Paraíba do Sul River Basin Agency

Created on June 20, 2002, AGEVAP or Pro Management Association of Waters in the Paraíba do Sul River Basin, was created to exercise the functions of executive secretary of CEIVAP, also developing the functions defined in art. 44 of Law No. 9,433 / 97, which deals with the competencies of the so-called Water Agencies, or Basin Agencies, as they are best known, especially with regard to the preparation of the Water Resources Plan and the execution of the actions deliberated by the Committee to the management of the basin's water resources (http://ceivap.org.br/agevap_1.php).

From the edition of Provisional Measure No. 165/04, later converted into Law No. 10,881 / 04, AGEVAP, after being qualified by the CNRH as a delegating entity of the Water Agency functions, may, after the establishment of a Management Contract with National Water Agency - ANA, to assume the functions of a Basin Agency, which are, essentially, to receive the resources from the charging for the use of raw water and invest them according to the Investment Plan, approved by the Basin Committee - CEIVAP.

AGEVAP has the legal personality of a private, non-profit association whose members are members of CEIVAP, which makes up its General Assembly. It is managed by the Board of Directors, whose members are appointed by the General Meeting, Fiscal Council and Executive Board, which is formed by a Director and two Coordinators.

2.5.2 Piracicaba-Capivari-Jundiaí River Basin Agency (PCJ)

With Law No. 9.433 / 97, which instituted the National Policy for Water Resources, it became possible to implement charging for the use of water, which is one of its instruments and aims to encourage rationalization for the use of water and generate resources for the application in projects aimed at the recovery of hydrographic basins. The second collection initiative in rivers in the domain of the Union is taking place within the scope of the PCJ basins (Piracicaba, Capivari and Jundiaí), specifically in the Piracicaba river and trainers (http://www.agenciadeaguapcj.org.br/).

In order to reach the implementation of charging for water use and implementation of the PCJ Water Agency, according to the relevant legislation, some steps were taken:

a) Creation of the Hydrographic Basin Committee: according to legal requirement, after approval by the CNRH, the PCJ Committee (Piracicaba, Capivari and Jundiaí) was created in March 2003, within the scope of the Union, involving 60 municipalities in São Paulo and four in Minas Gerais, considering only those whose headquarters belong to the same hydrographic region.

b) Basin Committee Resolutions: in order to allow the implementation of the collection, resolutions No. 024/05, which established the general criteria for the collection, including the amounts to be charged and No. 025/05, were approved by the basin committee, made it possible for the PCJ Consortium to sign a Management Contract with the National Water Agency (Agency of the Ministry of the Environment), allowing it to exercise the Water Agency functions for the PCJ Committee for a period of two years.

c) Approval of resolutions at the Federal level: the afore mentioned resolutions were approved by the National Council for Water Resources in November 2005, through resolutions CNRH nº 52 and 53, dated 11/28/2005, allowing the beginning of collection for 2006.

d) Registration / Update of Users: supporting this initiative, the National Water Agency (ANA), the Department of Water and Electricity (DAEE) of the State of São Paulo, and the Minas Gerais Water Management Institute (Igam) of the

State of In December 2005, Minas Gerais promoted a regularization process, which began with the registration or updating of the data of those who use the waters of rivers, reservoirs and lakes and concludes with the issuance of the granting of rights for these uses, including all urban sanitation service providers, industries, mining companies, aquaculture farmers and other rural uses, including those users who do not have a license, in order to recognize and organize the various uses to make the management of water resources in watersheds more efficient, in addition to allow the user to update his registration data in order to calculate the amounts charged.

e) Implementation of the PCJ Water Agency: After the approval of the collection and delegation to the PCJ Consortium, within the scope of the PCJ Committees and in the CNRH, the official inauguration of the Agency took place on 12/16/05, a technical and administrative structure was set up in municipality of Piracicaba-SP, which currently has three equipped rooms and, several results already achieved, within the goals of the Management Contract signed with ANA.

f) Revision of Collection Mechanisms: The joint resolution of the PCJ 024/05 committees provided for the revision of collection mechanisms, after the second year. Therefore, through Joint Deliberation 078/07, of 10/05/07, amended by Joint Deliberation 084/07, of 12/20/07. The review of the mechanisms was approved by the CNRH on 10/12/07 through Resolution 078/07.

g) Extension of the Deadline for Delegating Water Agency Functions to the PCJ Consortium: After the 2-year delegation period established in Joint Resolution PCJ 25/05, of 10/21/05, and CNRH Resolution No. 53, of 11/28/05, the PCJ Committees decided to extend the term of the delegation of water agency functions to the PCJ Consortium for another 4 years, through Joint Deliberation No. 080/07, of 10/05/07. The extension was also approved by the CNRH through Resolution 77, of 10/12/07.

When these processes involve the interaction of representatives of disconnected groups and the reallocation of competencies from existing state bodies, the construction of political-institutional capacity is crucial. This condition has been ignored by technical groups responsible for the design and implementation of the reform of the water management system. Without the gradual construction of social capital within the basin committees, actors who do not normally communicate will remain isolated" (Abers & Keck, 2004). The question of society's participation in these new decision arenas needs studies that can answer, according to Abers & Keck (2007), to the following points:

a) whether civil society can influence policies through these new deliberative bodies; b) whether it can hold the State responsible for its actions; c) whether these spaces genuinely represent the constituents on whose behalf they speak; d) whether the "politically excluded" are effectively represented; e) how democratic decisions are made internally.

3. Conclusions and Recommendations

The management of water resources in Brazil, as demonstrated in the previous chapters, has developed in a fragmented and centralized way. In addition, each user sector (electricity, navigation, irrigated agriculture, sanitation, etc.) carried out its planning and actions in an individualized and disconnected manner. Since the 1980s, there has been an increasing progress towards decentralization and the participation of civil society in the decision-making process.

It is understood, therefore, as basic principles of this new model of water resources management, the decentralized management in the scope of the hydrographic basin, the integration of the sectorial policies related to the water issue, the involvement of water users and civil society in decision making. decision, the treatment of water as an asset of economic value and not an inexhaustible gift from nature.

In this sense, numerous technical tools of management (instrumental rationality) and negotiation (dialogical rationality) have been used in the implementation of the various instruments provided for in Law 9.433 / 97. Such instruments, however, as demonstrated, must be the stage for constant improvement and incorporation of new scientific and technological advances.

It is noticed, meanwhile, that there must be a greater integration with the Industrial Policy and the Sectorial Policies, as well as with the Economic Development Council (CADE) and the consumer protection agencies. As Paulo Affonso Leme Machado (2001) emphasizes, "so that participatory management is not destroyed and that it does not become ineffective, social control must find ways of continuous and organized information". Therefore, an increase in the participation of civil society should be sought, in quantity and quality, in the Basin Committees, State Councils and the National Council of Water Resources - CNRH.

Murilo Ramos (2003) adds that "the existing regulatory agencies need to improve, and greatly improve, their mechanisms of relationship with society, especially consultations, hearings and public sessions, today much more formal processes of self-justification than constant interaction and effective with individuals, entities and associations that have no direct economic interest in the regulated area".

Another fundamental aspect for the improvement of the Brazilian regulatory framework in the opinion of Alberto dos Santos (2003) is:

> "The expansion of accountability mechanisms. An example of the inefficiency of these mechanisms can be exemplified in the public consultation process. In most cases, agencies respond to inquiries without proper reasoning or simply do not respond to inquiries. Thus, in order to have greater legitimacy, regulatory bodies must be more transparent and accessible to the controls of Parliament, the Executive Branch and society, with the effective implementation of instruments available in the legislation, proposed by the ministries and idealized by scholars on the topic".

According to Torres (2007), "the developments of Olson's theory are manifested more vehemently in historically *patrimonialist* societies such as the Brazilian one. In Brazil, unlike the United States, the State created society, spreading and consolidating paternalistic, nepotistic and oligarchic characteristics. History has shown that the counterpart of creating a strong, ubiquitous and centralizing patrimonial state, a direct inheritance of Iberian colonization, is the emergence of a weak, disorganized and apathetic civil society".

Barbosa Gomes (2003) argues that agencies can be created both at the federal and state levels, with the objective of regulating the provision by private operators of public services delegated to the private sector. However, two models have been followed: on the one hand, the "specialized sector model", in which several agencies are created, one for each sector (as in the case of federal agencies); and the "multisectoral model", in which only one agency is created to regulate all or almost all public services provided by private individuals, as is the case with most delegated agencies created in the states. However, greater articulation and integration between these entities is necessary. Thus, as Mattos (2006) points out,

> "Therefore, in the Brazilian legal-institutional experience, there would be no understanding of democracy as a regulatory principle in the process of formulating economic policies and regulating markets. There would be a lack of a positive notion of the State combined with a positive notion of democracy in the legal-institutional experiences of State intervention in the economy. Could the furtadian experience - theoretically and politically - have been the harbinger of this combination? Perhaps, but unfortunately it was not consolidated on the theoretical level and had a short life on the political level".

According to Jessé de Souza (2009), in Brazil, where an old academic debate still imagines the country dominated by "*jeitinho*" or "pre-modern personal relationships" - as if here a mysterious "evil of origin" had prevented that market and State (only in Brazil, among all countries on the globe) failed to develop the virtualities of a modern and impersonal society - Habermasian theory can be richly used. In a "Habermasian" register, Brazilian social problems appear to stem from an almost absolute "colonization" of market interests and money over all other social spheres. As a critical public sphere did not develop here - except episodically - neither did moral and political consensus develop, capable of opposing the simple indiscriminate use of everything and everyone as the goal of profit. The permanence and naturalization of the abysmal Brazilian social inequality in all dimensions comes, therefore, not from a corrupt "way" that is always of the State (demonized) and never of the market (deified) - as in the vision of our dominant liberalism that is conservative and pseudocritical - but the lack of self-criticism that permeates the whole of society. Critical thinkers, like Habermas, should not be used only as a means of "learned distinction", as a mere "ornament of intelligence", as an objective in itself, as is so common among us. They are a "practical weapon" to perceive our own society in another way, more critical and less self-indulgent and superficial as we are used to perceiving ourselves.

With regards to the training of civil servants and civil society for the effective management of water resources, it is worth highlighting the aspects of the Harbermasian theory, brought by Bannel (2009),

> "Perhaps the greatest challenge of Habermasian thinking for education today is to understand the educational process as the simultaneous formation of the individual as an irreplaceable individual, with his personal identity and life project, also as a member of a group any social and cultural, with its cultural, ethnic, racial identity, etc., as well as a citizen, that is, a member of a larger political community, which includes different social groups. Such training focuses on the development of communicative competence, necessary for communicative action, the mechanism responsible for both the reproduction of cultural traditions, forms of knowledge and moral norms, as well as their transformation".

It is hoped, therefore, with this study, that a simple assistance was given in the task of making a diagnosis of the pathologies and potential emancipators of contemporary Brazilian society, with regard to the shared management of its water resources.

4. Bibliographic References

ABERS, R. & KECK, M.: Comitês de Bacia no Brasil: uma abordagem política no estudo da participação social. R. B. Estudos Urbanos e Regionais, v.6, n.1, maio 2004, pp. 55-68.

ABERS, R. N., JOHNSSON, R. M. F., FRANK, B., KECK, M. E. & LEMOS, M. C.: Stakeholder Councils and River Basin Management in Brazil: Democratizing Water Policy? Anais do III Encontro da ANNPAS, 23 a 26 de maio de 2006, Brasília – DF.

ABERS, R. & JORGE, K. D. Descentralização da Gestão da Água: por que os comitês de bacia estão sendo criados? Ambiente & Sociedade, vol. VIII, nº 2, jul/dez 2005.

ABERS, R. N. & KECK, M.: Mobilizing the State: The Erratic Partner in Brazil's Participatory Water Policy. In Annual Meeting of the American Political Science Association, Philadelphia, PA, August 31-September 3, 2006 (www.apsanet.org).

ABRUCIO, F. L.: Trajetória recente da gestão pública brasileira: um balanço crítico e a renovação da agenda de reformas. Rev. Adm. Pública, Rio de Janeiro, vol.41, nº. esp, 2007. pp. 67-86.

AGÊNCIA NACIONAL DE ÁGUAS. www.ana.gov.br

AGÊNCIA NACIONAL DE ÁGUAS. Prestação de Contas – 2005.

AGÊNCIA NACIONAL DE ÁGUAS. Prestação de Contas – 2006.

AGÊNCIA NACIONAL DE ÁGUAS. Prestação de Contas – 2007.

CHAPTER 2

Water regulation and participative water resource allocations mechanisms in Brazil

Marcos Airton de Sousa Freitas

Specialist in Water Resources at the National Water Agency - ANA; Ministry of Regional Development - MDR

Abstract: This study analyzes the current model of water resource management in Brazil, with its progress and problems of implementation, in order to submit offers to its improvement. Its aims to discuss the current model of water resources management, with emphasis on regulation and control (search for efficiency and quality improvement and recovery of the public sphere as an instrument of citizenship). To achieve the stated objectives, we sought to examine the issue from a theoretical-historical approach (models of government in place in Brazil and genesis of the regulatory agencies) and participatory management, involving the relationship between the state and the sphere public. It follows that the current model lacks a qualified and effective participation of civil society, which is extremely important for the improvement of water governance and deliberative democracy in Brazil.

Keywords: water regulation, water resource allocation, water resources management.

1. Introduction

In the last decades, several studies have analyzed the management and regulation of water resources in the light of economic, environmental, ethical and social aspects (Freitas & Billib, 1997; Freitas & Amorim, 1999; Christofidis, 2001; Boff, 2003; Lobato da Costa, 2003; Porto & Lobato da Costa, 2004; Freitas & Freitas, 2004; Freitas, 2009; Freitas, 2010; etc). The current situation points to a situation of crisis (Haddad, 2008). The crisis around water reflects the crisis of conscience of our civilization and of the current, unequal, exclusive and depleting and exhausting world model of natural resources. In the process of building the sustainable management model of water resources in vogue, the big challenge is to establish a shared and decentralized power relationship, creating an opportunity for social participation, building consensus, settling conflicts and agreeing on unity in diversity (MMA, 2008).

Some experiences in different hydrographic basins, carried out through the implementation of the instruments of the National Water Resources Policy (granting of water resources rights, water basin master plans, charging for the use of water resources, classification of water bodies according to the predominant use etc.), as well as inspection campaigns, training programs, financing and stimulating research, the community organization has been adopted in Brazil. However, little has been researched in order to verify the real impacts and effects of these plans, programs and projects, especially those related to the "compatibility" of the different uses of water, those of improving the quantity and quality of water in hydrographic basins. Brazilian, participation and training of users and civil society for decision making within the scope of consultative and deliberative bodies (basin committees, water agencies, etc.). Research related, in particular, to the efficiency and effectiveness of the instruments and regulatory mechanisms advocated in Law No. 9.433 / 97, such as the collection and regulatory role of the National Water Agency - ANA and the Basin Agencies, are rarely found in literature.

2. Objectives

This chapter therefore intends to deal with the analysis of the current model of regulation and management of water resources, with its advances and implementation problems, with a view to presenting proposals for its improvement. Therefore, its general objective is to discuss the current model in the light of several aspects, in particular, social participation and the rescue of the public sphere and social representation (Habermas, 1981 and 1990; Offe, 1994; Cohen & Arato, 1992; Avritzer, 2003; Bordenave, 1995; Grau, 1998; Lubenow, 2007; Honneth, 2007).

Thus, its specific objectives are as follows: i) to approach the water resource management models put into practice in Brazil, contextualizing them from the administrative models in force in each era (classic Weberian bureaucratic model of the 1930s of the last century; model systematized in Decree-Law 200/67; and Managerial Reform of Brazilian Public Administration, from 1995 onwards; and the decentralized and participatory systemic model, resulting from the Federal Constitution, 1988 and Law No. 9,433 / 97); ii) analyze the current model of water resources management, with emphasis on aspects of regulation and social control (search for efficiency, improvement of quality and rescue of the public sphere as an instrument for exercising citizenship); iii) among the various instances that make up the so-called SINGREH - National Water Resources Management System, focus the analysis on basin committees (aspects related to social participation: participation in councils) and on the two Water Agencies already implemented, which are: Paraíba do Sul and PCJ (regulatory aspects: administrative autonomy; management contracts; valuing the careers of agency employees; decision-making models, etc.); iv) propose actions and proposals that contribute to the improvement of water resources management.

3. Water Resources Regulation

In order to achieve the above objectives, the theme will be analyzed from a theoretical-historical approach to the regulation and participatory management of water resources, supported mainly by the studies of Jürgen Habermas and Alex Honneth, seeking to contemplate the aspects of the State (public management and regulatory state), the market (regulation and management of water resources in Brazil) and the public sphere (public sphere in the contemporary world).

In this chapter, the spheres "State" and "market" (corporate modernization) will be addressed, within an instrumental logic, as well as the *Lebenswelt* sphere or "world of life" (cultural modernity), within a logical view of communicative action. Therefore, themes such as institutional independence and transparency in public management are considered, as well as the question of the public sphere and how social control has been considered in the management of water resources. The most recent transformations, therefore, are discussed, emphasizing the processes of differentiation (*Ausdifferenzierung*), rationalization (*Rationalisierung*), autonomization (*Autonomisierung*) and dissociation (*Entkoppelung*), which took place within the scope of water resources management and regulation in contemporary Brazil. In this context, it will be visualized how it has been happening, not only the contamination of the subsystems (economy and State) by rationalization, but also the expansion of that to some institutions in the world of life, through its colonization (*Kolonialisierung*). Finally, how could a paradigm shift be made within water management: from instrumental to communicative action, from subjectivity to intersubjectivity and from monological reason to dialogic reason. Finally, the main conclusions and recommendations of this study are presented.

The state regulatory function has its provision expressed in art. 174, of the Federal Constitution, of 1988. However, permeating the changes described above, so-called regulatory agencies appear, including the National Water Agency - ANA. Regulation, as Barbosa Gomes (2003) points out, represents "a kind of corrective indispensable to two processes that are intertwined: the deformation of the capitalist regime and the way in which the State engendered by that same regime". He further says that "from an absentee state and a mere guarantor of order and compliance with contracts, the maximum expression of property rights, the world has witnessed the emergence of an interventionist state, a provider of benefits aimed at minimizing and correcting imperfections and iniquities of the capitalist system".

While in the USA, regulation through regulatory agencies represented the rupture of the concept of the minimum state and the abandonment of the view of the state regulated by the invisible hand of the market, in Brazil, according to Barbosa Gomes (2003), attempts are made to "abandon a conception of a highly *clientelistic* state, but not to regulate it effectively, but to serve the interests of the various higher estates to which it has always been imprisoned ". Explains Barbosa Gomes (2003), with some dose of exaggeration, that it is "an implant, a *greffe* applied to fabrics of different texture. In short, another attempt to give the same remedy to symptoms and patients with totally different diagnoses".

> Barbosa Gomes (2003) clarifies that agencies can be created, both at the federal and state levels, with the objective of regulating the provision by private operators of public services delegated to the private initiative. The reproduction of this regulatory trend has followed two models: the "specialized sectorial model", in which several agencies are created, one for each sector (as in the case of federal agencies); and the "multisectoral model", in which only one agency is created to regulate all or almost all public services provided by private individuals, as is the case with most delegated agencies created in the states.

Gelis Filho (2006) defines regulation as "the intervention of the State in the economy and in social activity with the purpose of correcting market failures and increasing welfare, without this intervention implying the direct production of goods and services by state institutions". Melo (2000: 56) *apud* Gelis Filho (2006: 593) identifies four modal types of state regulatory action in the economy: public ownership of firms or entire sectors of the economy; exercise of regulatory activities directly by departments or agencies of the executive bureaucracy; various forms of self-regulation through corporate arrangements; and public regulation with a private property regime. Thus, Gelis Filho (2006: 595) considers that the possible forms of state regulatory intervention in the economy are variations on two basic possibilities: stimuli in relation to prices or control of the agents' behavior in a direct way.

The increase in demand, coupled with the scarcity and deterioration of the quality of water resources, cause serious conflicts to the multiple use of water, requiring new management paradigms. The Federal Law nº 9.433, of 08.01.1997, which instituted the National Water Resources Policy, has as basic principles, among others, the recognition of water as a vulnerable, finite resource with economic value and the decentralized and participative management of resources water resources. The water management models, that is, the institutional (legal and organizational) and financial mechanisms, have evolved over time in three distinct phases, namely: i) the bureaucratic model; ii) the economic-financial model; and iii) the systemic model of participatory integration. Such models have a close relationship with the models of administration of organizations and with the concept of society of Habermas Theory of Modernity, composed of two "worlds": the "system" (State and market) and the "world of life". Next, the main characteristics of each model will be analyzed, as well as the advantages, disadvantages and problems arising from their implementation in the water resources area (Freitas & Freitas, 1998).

Habermas (1981) defends a theory that would make it possible to resolve the conflicts in force in society and, not with a simple solution, but the best solution - one that is the result of the consent of all interested parties. Its greatest relevance is, undoubtedly, in seeking an end to arbitrariness and coercion in the issues that surround the whole community, proposing a way to have a more active and equal participation of all citizens in the disputes that involve them and, concomitantly, obtain the so longed for justice. This form defended by Habermas is communicative action (theory of communicative action), which branches into communicative action and discourse.

After the 1988 Federal Constitution, the ideas of decentralization and participation took on a new meaning in the Brazilian political-administrative arena, becoming important issues for governments that have sustained the hegemonic point of view on the modernization of the State. These processes of institutional and social changes, introduced through government policies, occur, however, in an extremely varied manner, in view, among other aspects, of the great political-administrative, environmental, social heterogeneity of the different Brazilian hydrographic regions.

Public administration has undergone numerous recent initiatives, in the sense of becoming increasingly transparent, more visible and closer to the population, through the implementation of various instruments of participation and control. According to Torres (2007), a set of factors would explain, in part, this phenomenon: i) the consolidation of Brazilian democracy since 1985, with the improvement of the checks and balances systems between the Executive, Legislative and Judiciary; ii) the active and comprehensive role of the Public Prosecution Service, due to the new attributions achieved in CF / 88; iii) freedom of the press, which enabled the development of more independent and investigative media; iv) the experiences of participatory budgeting, due to their pedagogical and mobilizing character; v) advances in legislation, such as

Complementary Law No. 101, of May 4, 2000, also known as the Fiscal Responsibility Law; vi) the gains in transparency provided by the enormous advances in electronic governance, as is the classic case of the use of the new type of bidding called the auction.

Supporters of the current water resource management model advocate that with the implementation of charging, a virtuous circle would trigger in the hydrographic basin. On the one hand, charging would induce the rationalization of water use, reducing its consumption and there would be less discharge of polluting effluents into bodies of water. On the other hand, the collection would generate funds for investments in the protection and recovery of waters in the basin, since almost all water laws provide for the preferential use of resources collected in the same basin where they were collected. In addition, these resources would encourage collaboration between municipal and state agencies, allowing the search for technical solutions not implemented due to budget restrictions. Thus, charging would be the catalyst for collaborative governance.

However, Abers & Keck (2004) point out that, first, the local economy of most of the Brazilian basins is not dynamic enough so that the amount collected by the collection is compatible with the level of investments necessary for the recovery of its waters. Thus, according to the authors, "these basins would remain dependent on other sources of financing, which would certainly restrict the decision-making capacity of the committees. Therefore, collection resources alone would not be sufficient to encourage stakeholder participation and eventual collaboration. Second, in richer basins, the implementation of collection tends to mobilize antagonistic civil interests and small users, and experienced in influencing decision-making processes, these groups would have the means to "capture" committees and agencies, with the aim of boycotting the collection or ensure that the system that they also finance mainly meets their needs". Abers & Keck (2004) affirm that this type of "capture" is already visible in the Paraíba do Sul river basin. And that this would "contradict the vision of governance by stakeholders, which assumes that different interests must have the same opportunity to influence policy. The chance of capture by economically more influential groups is especially strong in the case of the Paraíba do Sul basin, as it is one of the most dynamic and industrialized regions in Brazil, where economic interests are powerful and organized".

Abers & Keck (2004) question, what would happen, then, in basins with low collection potential. What would make committees work effectively in the absence of billing? If charging were to be applied widely, how to minimize the possibility of a "capture" scenario - or how to reverse it? Creating committees capable of intervening in water management involves not only the "internal policy" of creating an agenda and negotiating, but also the "foreign policy" of gaining support from relevant institutions. And they reiterate, saying "Brazilian technicians, especially in the highly insulated world of water resource management, seem to perceive political dynamics in a simplistic way, with something that hinders the technical rationality of technical solutions to problems".

They conclude by stating that "new institutional and decision-making frameworks can be created by legal mandate, but their implementation and operation involve socio-political processes, through which individuals and organizations face the task of making legal mandates a reality. When these processes involve the interaction of representatives of disconnected groups and the reallocation of competencies from existing state bodies, the construction of political-institutional capacity is crucial. This condition has been ignored by technical groups responsible for the design and implementation of the reform of the water management system. Without the gradual construction of social capital within the basin committees, actors who do not normally communicate will remain isolated" (Abers & Keck, 2004).

The question of society's participation in these new decision arenas needs studies that can answer, according to Abers & Keck (2007), to the following points: a) if civil society can influence policies through these new deliberative bodies; b) whether it can hold the State responsible for its actions; c) whether these spaces genuinely represent the constituents on whose behalf they speak; d) whether the "politically excluded" are effectively represented; e) how democratic decisions are made internally.

4. Conclusions and Recommendations

The management of water resources in Brazil, as demonstrated in the previous chapters, has developed in a fragmented and centralized way. In addition, each user sector (electricity, navigation, irrigated agriculture, sanitation, etc.) carried out its planning and actions in an individualized and disconnected manner. Since the 1980s, there has been an increasing progress towards decentralization and the participation of civil society in the decision-making process.

It is understood, therefore, as basic principles of this new model of water resources management, the decentralized management in the scope of the hydrographic basin, the integration of the sectorial policies related to the water question, the involvement of water users and civil society in decision making. decision, the treatment of water as an asset of economic value and not an inexhaustible gift from nature.

In this sense, numerous technical tools of management (instrumental rationality) and negotiation (dialogical rationality) have been used in the implementation of the various instruments provided for in Law 9.433 / 97. Such instruments, however, as demonstrated, must be the stage for constant improvement and incorporation of new scientific and technological advances.

It is clear, meanwhile, that there must be greater integration with Industrial Policy and Sectorial Policies, as well as with the Economic Development Council (CADE in Portuguese) and consumer protection agencies. As Paulo Affonso Leme Machado (2001) emphasizes, "so that participatory management is not destroyed and that it does not become ineffective, social control must find ways of continuous and organized information". Therefore, an increase in the participation of civil society should be sought, in quantity and quality, in the Basin Committees, State Councils and the National Council of Water Resources - CNRH.

Murilo Ramos (2003) adds that "the existing regulatory agencies need to improve, and greatly improve, their mechanisms of relationship with society, especially consultations, hearings and public sessions, today much more formal processes of self-justification than constant interaction and effective with individuals, entities and associations that have no direct economic interest in the regulated area".

Another fundamental aspect for the improvement of the Brazilian regulatory framework in the opinion of Alberto dos Santos (2003) is:

> "The expansion of accountability mechanisms. An example of the inefficiency of these mechanisms can be exemplified in the public consultation process. In most cases, agencies respond to inquiries without proper reasoning or simply do not respond to inquiries. Thus, in order to have greater legitimacy, regulatory bodies must be more transparent and accessible to the controls of Parliament, the Executive Branch and society, with the effective implementation of instruments available in the legislation, proposed by the ministries and idealized by scholars on the topic".

According to Torres (2007), "the developments of Olson's theory are manifested more vehemently in historically *patrimonialist* societies such as the Brazilian one. In Brazil, unlike the United States, the State created society, spreading and consolidating paternalistic, nepotistic and oligarchic characteristics. History has shown that the counterpart of creating a strong, ubiquitous and centralizing patrimonial state, a direct inheritance of Iberian colonization, is the emergence of a weak, disorganized and apathetic civil society".

Barbosa Gomes (2003) argues that agencies can be created both at the federal and state levels, with the objective of regulating the provision by private operators of public services delegated to the private sector. However, two models have been followed: on the one hand, the "specialized sector model", in which several agencies are created, one for each sector (as in the case of federal agencies); and the "multisectoral model", in which only one agency is created to regulate all or almost all public services provided by private individuals, as is the case with most delegated agencies created in the states. However, greater articulation and integration between these entities is necessary.

Thus, as Mattos (2006) points out,

> "Therefore, in the Brazilian legal-institutional experience, there would be no understanding of democracy as a regulatory principle in the process of formulating economic policies and regulating markets. There would be a lack of a positive notion of the State combined with a positive notion of democracy in the legal-institutional experiences of State intervention in the economy. Could the furtadian experience - theoretically and politically - have been the harbinger of this combination? Perhaps, but unfortunately it was not consolidated on the theoretical level and had a short life on the political level".

According to Jessé de Souza (2009), in Brazil, where an old academic debate still imagines the country dominated by *"jeitinho"* or by "pre-modern personal relationships" - as if here a mysterious "evil of origin" had prevented that market and State (only in Brazil, among all countries on the globe) failed to develop the virtualities of a modern and impersonal society - Habermasian theory can be richly used. In a "Habermasian" register, Brazilian social problems seem to stem from an almost absolute "colonization" of market interests and money over all other social spheres. As a critical public sphere did not develop here - except episodically - neither did moral and political consensus develop, capable of opposing the simple indiscriminate use of everything and everyone as the goal of profit. The permanence and naturalization of the abysmal Brazilian social inequality in all dimensions comes, therefore, not from a corrupt "way" that is always of the State (demonized) and never of the market (deified) - as in the vision of our dominant liberalism that is conservative and pseudocritical - but the lack of self-criticism that permeates the whole of society.

Critical thinkers, like Habermas, should not be used only as a means of "scholarly distinction", as a mere "adornment of intelligence", as an end in itself, as is so common among us. They are a "practical weapon" for perceiving our own society in another way, more critical and less self-indulgent and superficial as we are used to perceiving ourselves.

With regards to the training of civil servants and civil society for the effective management of water resources, it is worth highlighting the aspects of the Harbermasian theory, brought by Bannel (2009),

> "Perhaps the biggest challenge of Habermasian thinking for education today is to understand the educational process as the simultaneous formation of the individual as an irreplaceable individual, with his personal identity and life project, also as a member of a group any social and cultural, with its cultural, ethnic, racial identity, etc., as well as a citizen, that is, a member of a larger political community, which includes different social groups. Such training focuses on the development of communicative competence, necessary for communicative action, the mechanism responsible for both the reproduction of cultural traditions, forms of knowledge and moral norms, as well as their transformation".

It is hoped, therefore, with this study, to have given a simple assistance in the task of making a diagnosis of the pathologies and emancipatory potentials of contemporary Brazilian society, with regard to the shared management of its water resources.

5. References

ABERS, R. & KECK, M.: Comitês de Bacia no Brasil: uma abordagem política no estudo da participação social. R. B. Estudos Urbanos e Regionais, v.6, n.1, maio 2004, pp. 55-68.

ABERS, R. N. & KECK, M.: Mobilizing the State: The Erratic Partner in Brazil's Participatory Water Policy. In Annual Meeting of the American Political Science Association, Philadelphia, PA, August 31-September 3, 2006 (www.apsanet.org).

AVRITZER, L.: Governo Lula e o desafio da participação. Revista Teoria e Debate. Fundação Perseu Abramo, 2003.

BANNEL, R. I. Habermas e a educação. Dossiê Jürgen Habermas. Revista Cult, nº 136, 49-52p., 2009.

BARBOSA GOMES, Joaquim Benedito. Agências Reguladoras: A "metamorfose" do Estado e da Democracia. Disponível em http://www.mundojuridico.adv.br/documentos/artigos/texto027.doc, em junho de 2003.

BOFF, L.: Ética e gestão das águas. Palestra proferida no Seminário "Água, Desenvolvimento e Justiça Ambiental", Brasília, MMA, 2003.

BORDENAVE, J. E. D.: O que é participação. Coleção Primeiros Passos. 6ª ed., Brasiliense, 1995.

CHRISTOFIDIS, D.: Olhares sobre a política de recursos hídricos no Brasil: o caso da bacia do rio São Francisco. Tese de doutorado, Centro de Desenvolvimento Sustentável, Brasília, UnB, 2001.

COHEN, J.L. & ARATO, A.: Civil Society and Political Theory. Cambridge: The MIT Press, 1992.

FREITAS, M. A. S.: A Regulação dos Recursos Hídricos: estado e esfera pública na Gestão de recursos hídricos, Rio de Janeiro: Ed. CBJE, Rio de Janeiro, 174p., 2009

FREITAS, M. A. S.: Que venha a seca: modelos para a gestão de recursos hídricos em regiões semiáridas, Rio de Janeiro: Ed. CBJE, Rio de Janeiro, 413p., 2010.

FREITAS, M. A. S.; AMORIM, A. D.: Gestão das Águas: Um Desafio para o Piauí. Revista do CREA-PI, Teresina, v. 15, n. 6, p. 3-5, 1999.

FREITAS, M. A. S.; BILLIB, M. H. A.: Drought Prediction and Characteristic Analysis in Semi-arid Ceará-Northeast Brazil. In: Rosbjerg, D., Boutayeb, N.-E., Gustard, A., Kundzewicz, Z. W. and Rasmussen, P. F.. (Org.). Sustainability of Water Resources under Increasing Uncertainty. 1 ed. Rabat: IAHS, 1997, v. 240, p. 105-115.

FREITAS, M. A. S.; FREITAS, M. A. S. R.: Subsídios à Gestão Hidroambiental no Estado do Piauí. In: VII Simpósio de Recursos Hídricos do Nordeste, 2004, São Luis. Anais do VII Simpósio de Recursos Hídricos do Nordeste. Porto Alegre: Editora da ABRH, 2004. v. 1. p. 1-1.

FREITAS, M.A.S.; FREITAS, S.H.B.: Discussão acerca dos modelos de gestão de recursos hídricos. In: IV Encontro de Iniciação à Pesquisa, Fortaleza. Anais do IV Encontro de Iniciação à Pesquisa. Fortaleza: Editora da UNIFOR, 1998. v. 1. p. 64-67., 1998.

GELIS FILHO, A.: Análise comparativa do desenho normativo de instituições reguladoras do presente e do passado, Revista da Administração Pública – RAP, Rio de Janeiro, 40(4): 589-613, Jul./Ago. 2006.

GRAU, N. C.: Repensando o público através da sociedade – novas formas de gestão pública e representação social, Ed. Revan-ENAP, 1998.

HABERMAS, J.: Theorie des Kommunikativen Handelns. 2 v. Suhrkamp, 1981.

HABERMAS, J.: Strukturwandel der Öffentlichkeit. Suhrkamp. (Vorwort zur Neuauflage), 1990.

HABERMAS, J.: Kommunikatives Handeln und detranszendentalisierte Vernunft, PhilippReclam jun. GmbH & Co., Stuttgart, 2001.

HADDAD, P.R.: 2050 – petróleo ou água? O Estado de São Paulo, em 13/05/2008.

HONNETH, A. Sentimento de Indeterminação: uma reatualização da Filosofia do Direito de Hegel. Trad: Rúrion Soares Melo. São Paulo: Editora Singular, Esfera Pública, 2007.

JESSÉ DE SOUZA. Ambivalência moral e política do mundo moderno. Dossiê Jürgen Habermas. Revista Cult, nº 136, 60-62p., 2009.

LOBATO DA COSTA, F. J.: Estratégias para o gerenciamento dos recursos hídricos no Brasil: Áreas de cooperação com o Banco Mundial. Brasília-DF: BIRD, abr. de 2003.

LUBENOW, J.A.: A categoria de esfera pública em Jürgen Habermas: para uma reconstrução da autocrítica. Cadernos de Ética e Filosofia Política 10, 1/2007, p.103-123.

MATTOS, P. T. L.: A formação do estado regulador, Novos Estudos 76, novembro de 2006.

MINISTÉRIO DO MEIO AMBIENTE: Água: Manual de Uso – Vamos cuidar de nossas águas – Implementando o Plano Nacional de Recursos Hídricos, 2ª edição, Brasília – DF, 2008.

OFFE, C.: Contradicciones en el Estado del Bienestar. Alianza Universidad, Ciencias Sociales, Madrid, 1994.

PORTO, M. & LOBATO DA COSTA, F.J.: Mecanismos Econômicos, Sociais e Ambientais de Gestão da Água, Revista Rega, vol. 1, n.2, jul.-dez. 2004.

TORRES, M. D. F. Agências, Contratos e Oscips: a experiência pública brasileira, Rio de Janeiro: Editora FGV, 2007, 170p.

CHAPTER 3

A Resilient Drought Risk Management Approach in the Semiarid Northeast Brazil

Marcos Airton de Sousa Freitas[1], Paulo Breno de Moraes Silveira[2] & Gabriel Belmino Freitas[3]

Abstract: Droughts in Northeast Brazil, which tend to intensify due to climate change, have repeatedly brought famine, mass migration and social conflicts in this region. Its prediction, monitoring and management, however, remain a central research theme. In water resources management in semiarid regions such as the Northeast of Brazil, it is fundamental to have tools to aid decision making. This paper presents three components of the so-called SIGES (Drought Management System), the items related to drought prediction and monitoring, as well as many reservoir operation methodologies for water scarcity situations. Statistical models, artificial neural networks and machine learning techniques were used for drought prediction. In order to perform precipitation monitoring, several indexes were adapted and incorporated into a droughts basic characteristic monitoring system (duration, severity and intensity), so that different mitigating actions could be implemented in accordance with the values reached by these parameters. We utilized the following meteorological indexes for this purpose: Rainfall Anomaly Index (RAI); Bhalme & Mooley Drought Index (BMDI); Lamb Rainfall Departure Index (LRDI). Finally, some reservoir operation for water scarcity situations methodologies are presented and discussed. The described components were applied to the Northeast of Brazil, especially Piauí, Ceará and Rio Grande do Norte states.

Keywords: drought prediction, drought monitoring, reservoir operation optimization.

[1] Senior water resources specialist with National Water Agency (ANA); Address code: 70.610-200; E-mail: masfreitas@ana.gov.br
[2] Senior water resources specialist with National Water Agency (ANA); Address: SPO, Quadra 3, Lote 5, Bloco L, Brasília, DF, Brazil. E-mail: paulo.breno@ana.gov.br
[3] Economist from the University of Brasília (UnB); Campus Darcy Ribeiro s/n – Asa Norte, Brasília – DF – Brazil. Address code: 70.910-900; Master of Science student in Applied Economics at Leopold-Franzens Universität Innsbruck. E-mail: gabrielbelminofreitas@gmail.com

1. Introduction

Marengo (2006), Salati et al. (2007), Brito et al. (2017) conducted studies on the impacts of global climate change for many areas of Brazil, as Brazilian Amazon, Northeast, Pantanal and the Prata River Basin, showing precipitation and temperature anomalies, and water balance for the XXI century. The semiarid northeastern presenting short but crucially important rainy season in the current climate could, in a warmer climate in the future, become arid.

To the northeast, Salati et al. (2007) assessing climate variability in the region, showed that the average temperature increased by 0.6°C within the period 1991 to 2004, when referred to the period from 1961 to 1990. For maximum temperature values they indicated a 0.6°C increase and for minimum temperature values a 0.5ºC raise. The precipitation decreased 153 mm, a 11.6% drop.

Nobre et al. (2004) indicate that the future biomes distribution in South America may be affected by the combined impacts of climate and land use changes, which can take the system to the *savannization* of parts of the Amazon and the desertification of Northeast Brazil.

In relation to the observed values, the ensemble model mean values tend to be large in much of Africa, the Northeastern region of South America (northeastern Brazil), and northwest North America, and small in northern low latitudes of the Americas and southern South America. The arithmetic mean streamflow of 12 IPCC models for the period 2041-2060, when compared to the 1900-70 period A1B scenario for the rivers of northeastern Brazil showed a 15% reduction (Milly et al., 2005).

Several studies have indicated the influence of numerous atmospheric phenomena on rainfall in Northeast Brazil (Moura and Shukla, 1981; Hastenrath, 1984; Freitas, 1996; Freitas and Billib, 1997; Uvo et al., 1998; Andreoli and Kayano, 2007; Moscati and Gan, 2007). Also several climatological studies have indicated the existence of a strong relationship between sea surface temperature distribution (SST - sea surface temperature) along the tropical Atlantic basin temperature and the semiarid northeastern Brazil precipitation, as well as a decadal trend associated with changes in the meridional position of the ITCZ - Intertropical Convergence Zone (Moura and Schukla, 1981; Rao and Hada, 1990; Billib and Freitas, 1997). These phenomena are indicative to be related to climate variability and extreme droughts and floods in the region.

Droughts can be characterized as a natural phenomenon sharply differenced from other natural catastrophes. Unlike other natural occurrences as floods, hurricanes and earthquakes, that start and end at sudden, being restricted into a small region, drought phenomena are used to have quite often a slow start, a long duration, and is generally spread out through a wide area (Freitas, 1997).

Drought is known as a recurrent phenomenon in semi-arid regions. The lasting effects of a drought in a particular region depend, however, not only on the duration and intensity of drought, but also of socio-economic and cultural rights of the affected population. The occurrence of drought in regions where the water demand is greater than water availability or where there is a large variability of supply water, almost always bring large scale consequences. Large irrigation projects and densely populated urban concentrations are subject to a huge vulnerability with regards to water supply. During drought periods there is also a significant decrease in hydroelectric power generation.

Freitas (1996) presented a methodology for integrated regional drought analysis, which briefly consists of the following topics: (1) definition of different types of drought, (2) forecasting and monitoring: (3) water resources management and optimization, (4) effects evaluation and (5) planning of mitigating actions. Freitas (1996), and Billib & Freitas (1997) demonstrated the feasibility of using prediction models for the dry northeastern Brazil: statistical-probabilistic models and models based on neural networks.

Drought periods forecasting and monitoring are particularly useful for Northeast Brazil due to, among many others, the following: (1) the existence of numerous irrigation projects implemented and being implanted along major rivers, (2) supply water of large cities is mostly dependent on direct runoff from rivers or, indirectly, on the volume accumulated in dams, (3) most agricultural crops depend only on the regularity of the rains, (4) the possibility of using groundwater is small as compared to that of surface water and (5) most of the region's energy production is based on hydropower (Freitas, 1997).

It is a notorious fact the extreme need for implementation of Decision Support Systems – DSS, like SIGES – Drought Management System, in order to provide the policy makers, i.e. the decision makers (politicians, state secretaries, coordinators, etc..) of the various government agencies (at federal, state and municipal level) of skilled and practical instrumental, especially those related to water resources management, for the prediction and monitoring of dry and wet periods in semi-arid regions.

2. Drought Prediction

The dry and wet years forecast is of vital importance for essentially agricultural semiarid regions (Billib and Freitas, 1997). The semi-arid region with an area of more than 1 million km^2, is characterized by a large temporal and spatial precipitation variability, resulting in a process of main watercourses intermittency. It is clear the urgency of implementing models in order to reduce these uncertainties. The model therefore includes the analysis of dynamic weather causes i.e. the analysis of the global circulation system, especially ENSO (El Niño - Southern Oscillation), as well as surface Atlantic Ocean temperature anomalies (Atlantic Dipole phenomenon).

Several methods have been applied during last century in order to predict droughts in northeastern Brazil. Attempts, for example, were performed in order to relate the number of sunspots and total rainfall. Correlations between various variables such as air pressure, temperature, sea surface temperature and other from distant regions and annual precipitation in the Northeast were analyzed at the beginning of the last century (Walker, 1928).

Figure 1 shows a simplified diagram proposed by Freitas (1997) for drought management. This management system subcomponent of the integrated regional drought analysis consists of statistical methods application (correlation analysis, contingency tables, principal component analysis etc.), artificial neural networks, fuzzy logic etc. for rainfall forecasting. Another possible method would be to forecast sea surface temperature (SST) 6 or even 12 months in advance and then the total rainfall in a region through simulation using a global atmospheric-oceanic circulation model. The latter path however requires greater computational cost.

ENSO (El Niño - Southern Oscillation)

Besides the influence of the Intertropical Convergence Zone position', among other things, several authors have reported possible connection between El Niño and Southern Oscillation and the behavior of rainfall in Northeast Brazil. Ropelewski and Halpert (1987) among others have addressed this relationship.

Figure 2 shows the alternations between wet and dry periods and years ENSO - El Niño / Southern Oscillation, for the Ceará state according to Rasmusson and Carpenter (1983). The rainfall rate used was LRDI (Lamb Rainfall Departure Index) that expressed the regional rainfall deviation related to the mean, in terms of standard deviations (Lamb et al. 1986).

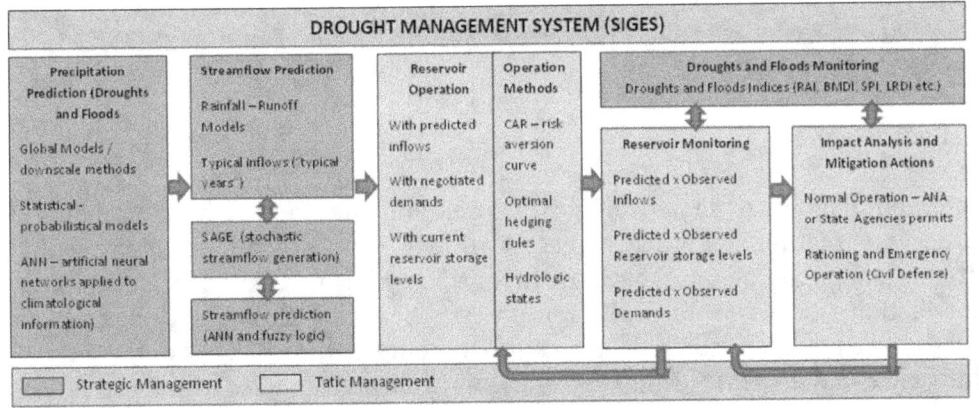

Figure 1: Drought Management System (SIGES)

It is easy to see that, generally drought years occur after El Niño phenomenon occurrence years. This happened in the years 1914, 1918, 1930, 1941, 1951, 1953, 1957, 1965, 1969, 1971, 1982, 1986, 1992 and 1997. The year following an El Niño year, however, is not always a dry year, as seen in 1912, 1924 and 1926. There are also dry years that did not follow El Niño years, such as 1936 and 1979.

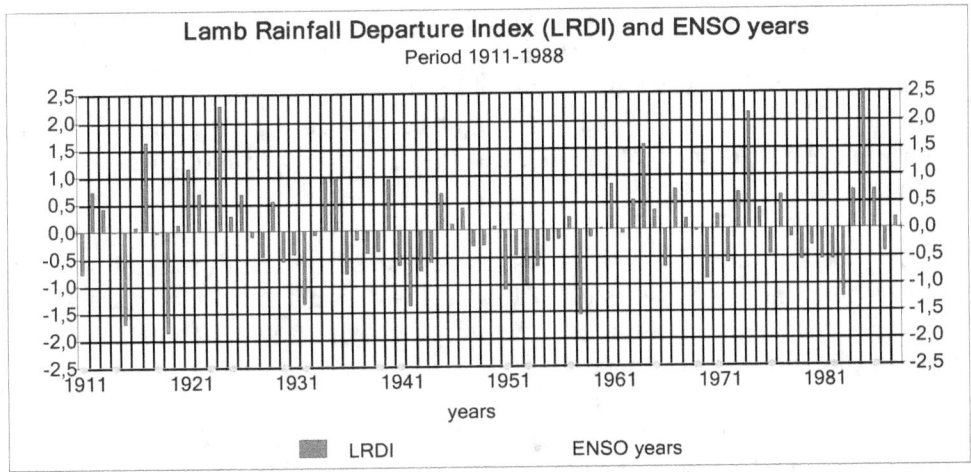

Figure 2: Alternation of dry and wet periods and ENSO years, from 1911 to 1988, in the State of Ceará (Northeastern Brazil), according to Rasmusson and Carpenter (1983).

El Niño is related to the Pacific sea surface overheating, nearby Peru and Equator coasts. South Oscillation, by its turn, is related to air pressure difference between Darwin (12º 20'S, 130º 52'E) and Tahiti (17º 33'S, 149º 31'W).

Statistical Models

The identification of meteorological causes and the development of drought forecasting methods in northeastern Brazil were made by analyzing the global circulation phenomena, especially the ENSO (El Niño-Southern Oscillation). Correlations between precipitation in Northeastern Brazil and circulation related parameters, such as sea surface temperature (SST) and pressure difference in the Pacific (Tahiti-Darwin), allow future rainfall estimates.

Regression analysis and the use of conditional probability tables were employed using rainfall data from the Northeast of Brazil, in addition to Pacific Ocean sea surface temperature and air pressure data. For the establishment of the model, the years were classified as, humid, normal and, dry, based on its probability of rainfall exceedance. Initial statistical analysis was based on thirty rainfall stations well spatially distributed in the Ceará state aiming to observe the dependence severity between El Niño incidence and droughts occurrence in Ceará. Additionally, we have established conditional probability tables for different stations. The 33% and 67% quantiles of the rainfall series were used to classify years in dry, normal or wet. ENSO Indexes were also classified into warm, normal or cool (Freitas, 1996). In this work, the data sources used were the SST and Pacific Ocean pressure differences presented by Wright (1989). These data were homogenized by Wright to take into accounting a possible change in the data density of the measurement methods or in the location of the station.

The sea surface temperature anomalies are based on the difference between those to the long-term average. Positive values of this anomaly are related to values above average. The data were multiplied by 100, so that a value of +120, for example, corresponds to a temperature of +1.2 ° C above the average.

The (Darwin - Tahiti) air pressure difference, hereafter DT, describes the pressure gradient between these Pacific endpoints and the so-called Southern Oscillation. During the "normal" years pressure at Darwin is smaller than the Tahiti, so DT has positive values meaning an inversion of the so-called Walker Circulation.

Regression Analysis and Conditioned Probability Tables

Various combinations of data for correlation coefficients were calculated for the studied rainfall stations, for monthly averages, three months, six months and annual totals. This was done to establish which was the best correlation coefficient data set.

For each station the rainfall exceedance probabilities corresponding to 33% and 67% were calculated'. These levels were used to classify the analyzed periods in dry, normal or wet. For SST anomalies limits -50 and +50, i.e. an anomaly of -0.5 ° C and +0.5 ° C, which also approximately correspond to 33% and 67% quantiles were chosen.

This analysis was carried out for rainfall average values of MAM(-1), JJA(-1) and SON(-1) and anomalies SSTs (DJF the following year), i.e. to 9 and 6 months advance, respectively. Figure 3 shows the results for the Mombaça station. The black dots indicate years when El Niño were from moderate to extreme, according to Rasmusson and Carpenter (1983) classification. The tables show conditioned probabilities for rainfall categories (wet, normal and dry) against SSTs anomalies (cool, normal, warm). For this station, for example, there is an anomaly greater than 0.5 ° C in MAM(-1) which denotes a rainfall probability of about 54% of presenting (DJF) below the 33% quantile (dry) or a probability of 92% to provide a total precipitated below 67% quantile.

For contingency diagrams and conditioned probabilities tables implementation on a forecast drought model using only Pacific Ocean sea temperature a check was performed in a forecast for Ceará state. In this test, probability tables of average values of all tables (stations) for the period of 9, 6 and 3 months before the period to be predicted (DJF) were implemented. Table 1 shows the results for the 6 months in advance period. It is assumed these tables are representative for the Ceará state.

Table 1 shows a 66% of a dry period (DJF) likelihood for SST indexes greater than +50 (i.e. +0,5ºC) In this circumstance, the probability associated with the dry and normal periods for a 6 month in advance forecast is found to be 88%.

Forecasts based on JJA (-1) SST show that dry periods in DJF in the Ceará State can be predicted nearly six months in advance. A total of twelve values at six stations employed for validating for the years 1983 and 1984, eight were estimated as dry, corresponding to about 67%, in a conditioned probability table. Likewise, eleven out of twelve refer to eight dry years and three normal years, many close to the value of 88% in the same table.

SSTs vs. Precipitation at Mombaça station

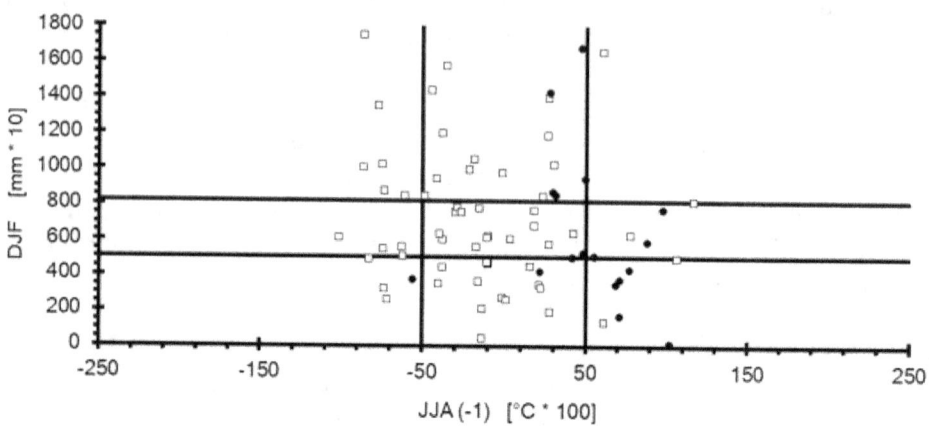

| JJA (-1) | cool SST | normal SST | warm SST |
DJF	(14 events)	(47 events)	(13 events)
Wet period	42 %	36 %	8 %
Normal period	29 %	34 %	38 %
Dry period	29 %	30 %	54 %

Warm SST :	x > 50;	Wet year :	y > 820
Normal SST :	-50 ≤ x ≤ 50;	Normal year :	500 ≤ y ≤ 820
Cool SST :	x < -50;	Dry year :	y < 500

ENSO years: 1911, 1914, 1918, 1923, 1925, 1930, 1932, 1939, 1941, 1951, 1953, 1957, 1965, 1969, 1971, 1976, 1982 (Rasmusson and Carpenter,1983)

Figure 3: Contingency Tables and Conditioned Probability Table for JJA(-1) SST vs Precipitation at Mombaça (Ceará State).

Table 1: Conditioned Probability Table for Ceará State: JJA(-1) SSTs vs. DJF-Precipitation

| JJA (-1) | cool SST | normal SST | warm SST |
DJF			

Wet period	31 %	39 %	12 %
Normal period	35 %	33 %	22 %
Dry period	34 %	28 %	66 %

Artificial Neural Networks and Machine Learning Techniques

Artificial Neural Networks provided a significant progress for systems theory and pattern recognition fields. Neural networks have a very flexible mathematical structure, being capable of non-linear relations identification and complex processes description. This name is given for models that replicate biological networks structures and performing, as the brain (Kosko, 1992). A neural network consists of a great number of elements, called neurons (cells, units), and a great number of connections (links), known as synapses. Each connection is associated with a weigh, which is intrinsically related to the network learning capacity. Thus, it is called an intelligent system, having, as a fundamental characteristic its learning capability.

Different network topologies and various input data combinations (time annual, quarterly and monthly, and Atlantic and Pacific Oceans data) were analyzed (Figure 4). The primary advantage of this methodology lies in the use of advanced mathematical techniques, which allow to achieve a given level of accuracy and advance an estimate of the total to be precipitated or drained within a given region during the rainy season. This prognosis is valuable for optimizing reservoir operation, especially when coupled with flow generation models (Billib and Freitas, 1996).

Two basic procedures were tested in each station: modeling their own rainfall series, and the use of SSTs Pacific and Atlantic oceans through neural networks. Then, the drought index was calculated for the thirty rainfall stations chosen in Ceará state: Lamb Rainfall Departure Index (LRDI) according to Lamb et al. (1986) for the rainy semester (January-June) and correlated with the Atlantic SST. Machine learning techniques have also been tested for drought index prediction and for streamflow prediction (Belayneh and Adamoswski, 2013; Raha and Gayen, 2019; Freitas, 2019).

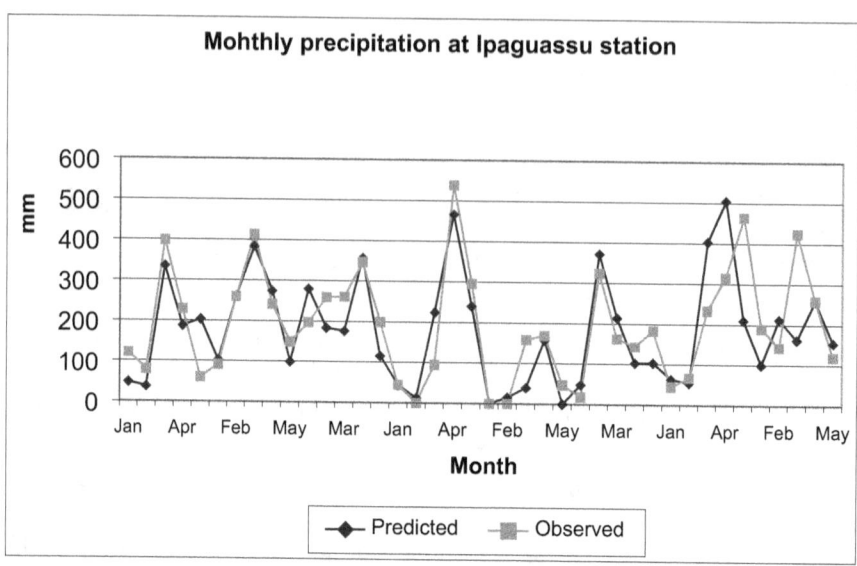

Figure 4: Predicted and observed monthly precipitation at Ipaguassu station (Ceará State), using the backpropagation through time algorithm, without adaptative process.

3. Drought Monitoring

Monitoring of drought periods can be done using indexes. Based on them, we can develop a drought characteristic monitoring system, as well as the different measures to be affected in accordance with the values of these parameters. The most known and widely used for the investigation of the spatial and temporal distribution of drought periods is the determination of some drought index. This index can usually be defined as a value which represents the cumulative effect caused by a long water deficit period (Freitas, 2005).

Due to the difficulty of having a single drought phenomenon definition, drought index determination is also problematic. Drought severity is usually expressed as a function of the mean value of one or several climate parameters. The most usual drought indexes express the total precipitation in given region as a known average data percentage. Thus, one can compare the current value with the average value of weekly, monthly, semesterly or annually series.

Drought indices

Transeau (1905) determined an effective precipitation index based on 150 rainfall stations data' in the United States, which expressed the relationship between precipitation and annual evaporation. Later, for the sake of simplification, the evaporation component was replaced by the temperature. An example is the so-called Lang Rain Factor (LRF). This index corresponds to the ratio between average annual precipitation and temperature. Another index, known as Martonne Aridity Index (MAI), is nothing more than a minor change to the above, given by the following equation: $N = MAI / (T +10)$. Subsequently, it was realized that the evapotranspiration instead of evaporation, was a decisive parameter in classifying the climate of a region. An index known too, which already takes into accounting the evaporation is called the Aridity Index (AI) formulated by Thornthwaite (1948), a measure of water deficit as a percentage of annual potential evapotranspiration. As the deviation from the normal condition, you can classify the AI into various classes.

Rainfall Anomaly Index (RAI)

The first index, however, that was incorporated into the Drought Management System - SIGES in order to make it possible to compare rainfall deviations related to the normal conditions of various regions is called the Rainfall Anomaly Index (RAI), described by Rooy (1965) and pioneered in Brazil by Freitas (2005).

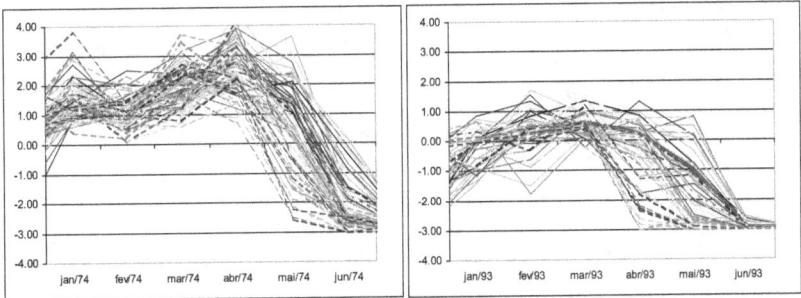

Figure 5: Rainfall Anomaly Index (RAI) for 185 Piauí State pluviometric stations, applied to a wet year (1974) and to a dry year (1993).

After this, numerous works showed its applicability in the Brazilian territory (Lima et al., 2016; Freitas, 2016; Martorano et al., 2017) and other countries (Raha and Gayen, 2019). In those studies, it was verified that the Rain Anomaly Index functioned as a good instrument for the study of seasonal precipitation.

Through this monitoring can generate predictions about the regional climatological variation.

Figure 5 shows an example of the application of this index to 185 rainfall stations in the Piauí state. It is easy to recognize 1974 as a wet year and 1993 as a typically dry year for almost all stations analyzed.

Bhalme & Mooley Drought Index (BMDI)

Palmer (1965) presented a water balance procedure, for the semi-arid region of western Kansas State and the sub-humid region of Iowa, USA later known as the Palmer Drought Severity Index (PDSI). The PDSI is calculated based on evapotranspiration, infiltration, runoff and other data, expressing a measure for the accumulated difference between the normal precipitation and the precipitation necessary for evaporation. This analysis is done on a weekly or monthly basis. This procedure gives out an index that ranges from -4 (extreme drought), through zero (normal) to +4 (very humid periods).

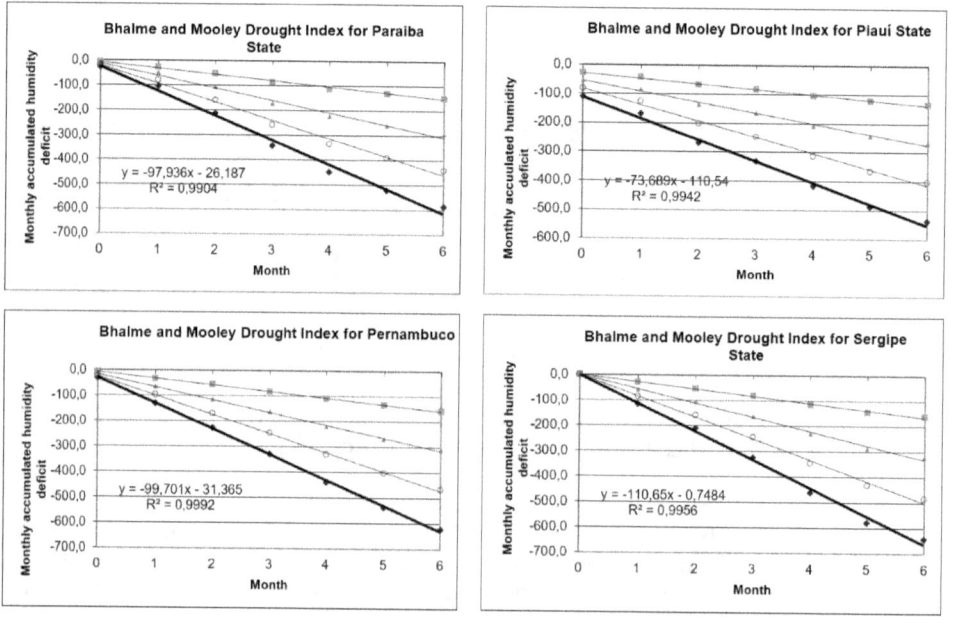

Figure 6: Bhalme & Mooley Drought Index (BMDI) for Paraiba, Piauí Pernambuco and Sergipe States.

Several authors (Alley, 1985; Guttman, 1991) demonstrated that the PDSI was not a good humid condition indicator, particularly during dry periods. Another

PDSI disadvantage is that the regularization of the flow surface is not considered. Bhalme and Mooley (1980) also pointed out the same problems for tropical regions of India. They then proposed a modification of the original content, in order to incorporate the climatic conditions prevailing in India. This index was known as Bhalme and Mooley Drought Index (BMDI). Figure 6 shows the application of this index for the Paraiba, Piauí, Pernambuco and Sergipe states. Freitas (2005) presented the application of BMDI for Ceará State and Freitas et al. (2012) showed its application for Rio Grande do Norte State.

Since this index present both positive and negative values it can be used in evaluating periods of droughts and floods. The values of current, monthly, accumulated during the crops growth or the rainy season (January to June) BMDI can then be compared with historical values in the region, so as to have a permanent control of the humid condition.

Lamb Rainfall Departure Index (LRDI)

The calculation of this ratio (Lamb et al., 1986) consists of a normalization procedure, through which average deviations of precipitation of various stations in given region are grouped to determine a unique index. A major advantage of this method is that all the precipitation series, which usually have many gaps, can be used in determining a regional index. Figure 2 shows the result of the use of this methodology to rainfall stations analyzed in the Ceará state.

4. Reservoir Operation Optimization using Climate Information

Among the various SIGES applications, we will report now those related to shortage impact mitigation based on forecasting and monitoring information for these periods. In semiarid regions with high evaporation rates the water resources manager often faces the problem of determining minimum water losses conditions for surface reservoir operation, considering different uses, their priorities shortages or rationing situations.

Silveira and Freitas (2004) deal with this considering two options: first, the model conventionally used, in which the total volume of water is assured for the most priority sectors while the lowest priority ones are exhausted. In the second, the quota for each user is divided into two parts: being exhausted first, the part that corresponds to the surplus of the balance of each economic sector, i.e., the region above its equilibrium point, moving then to ration the second part of the quota. Thus, an attempt is made to establish a criterion to minimize the possibility of breakage of entire sectors of the economy (optimal hedging rules and hydrological status methods). Another method for dealing with this problem is performed using the so-called Risk Aversion Curve (Gondim Filho et al., 2005).

In many hydrographic basins, due to the lack of fluviometric data, it is often necessary to obtain a series of affluent flows into the reservoir using rainfall-runoff or synthetic streamflow generation models. For the generation of synthetic flowrates in intermittent rivers typical of semiarid regions, Freitas (1995, 2010) presented the SAGE system - Stochastische AbflussGEnerierungsmodelle, consisting of the following models: i) PAR-Model (Thomas / Fiering) with modification of Clarke, ii) PAR-Model (Thomas / Fiering) with transformation Matalas, iii) Two-tier model (PAR (1) / AR (1) with log-gamma distribution), iv) Two-tier model (PAR (1) / AR (1) with log-normal distribution), v) Two-tier model (PAR (1) / GAR (1) of Fernandez & Salas; Fragment method-AR (1) log-gamma distribution, vi) Fragment method-AR (1) with log-normal distribution; vii) Fragment-GAR (1); viii) Disaggregation model / AR (1) by Valencia & Schaake. Later, Freitas (2010) presented applications of the model ARRF (Alternating Renewal Reward - Fragment) to various basins in northeastern Brazil.

5. Conclusions and Recommendations

The proposed models offer a practical approach for forecasting rainfall between 6 and 3 months in advance of the rainy season. Using the methods described above for predicting droughts, and various indexes of drought, combined with models of flow generation (Freitas, 1995), opens a new perspective in the management and optimization of water resources systems in semiarid regions, as the Northeast of Brazil.

The aim of the correlation analysis was to verify the correlation degree between El Niño and drought incidence in northeast Brazil. Data were initially grouped in various ways, for example, averaged for three (3) and six (6) months and later correlated with the index of ENSO, i.e., El Niño (SSTs) and Southern Oscillation (SOIs).

In northeastern Brazil, particularly in Piauí and Ceará states, extreme droughts often occur whenever there happens the phenomenon known as ENSO. The teleconnections between ENSO indices (Pacific) and interannual rainfall in these states can be used in drought forecast models, based on conditional probability tables and neural networks. These predictions once incorporated into rainfall-runoff models and streamflow generation models can contribute for the reservoir management and operation optimization in this region.

SIGES have been of great importance as a subside to mitigatory actions, specially the so-called no-regret actions for the current extreme drought period (2012-2019), as demonstrated on Abels et al. (2018).

6. Acknowledgements

This research was conducted partially with the financial support from the Bramar Project – Water Scarcity Mitigation in Northeast Brazil (German-Brazilian research and technology project sponsored by the Federal Ministry of Education and Research – FINEP: Ministério de Ciência, Tecnologia e Inovação).

7. References

Abels, A., Freitas, M. A. S., Pinnekamp, J., Rusteberg, B. 2018 *Bramar Project – Water Scarcity Mitigation in Northeast Brazil*, 1. ed. Aachen: Department of Environmental Engineering (ISA), 2018. v. 1. 155p.

Alley, W.M. 1985 The Palmer Drought Severity Index as a Measure of Hydrologic Drought, *Water Resources Bulletin*, **21**(1), 105-114.

Andreoli, R. V., Kayano, M. T. 2007 A Importância reativa do Atlântico Tropical Sul e Pacífico Leste na Variabilidade de Precipitação do Nordeste do Brasil, *Revista Brasileira de Meteorologia*, v.22, n.1, 63-74.

Belayneh, A., Adamoswski, J. 2013 Drought forecasting using new machine learning methods, Journal of Water and Land Development, n. 18(I-VI): 3-12.

Bhalme, H.N., Mooley, D.A. 1980 Large-Scale Drought/Floods and Monsoon Circulation, *Monthly Weather Review*, 108, 1197-1211.

Billib, M. H. A., Freitas, M. A. S. 1996 *Drought Forecasting and Management for Northeast-Brazil by Statistics, Neuro-fuzzy Systems Analysis and Stochastic Simulation*. In: Conference on Water Resources & Environment Research: towards the 21st Century, 1996, Kyoto.

Brito, S., Cunha, A. P., Castro, C., Alvala, R. Marengo, J., Carvalho, M. 2017. Frequency, duration and severity of drought in the Brazilian Semiarid region. International Journal of Climatology, doi: 10.1002/joc.5225.

Freitas, G. B. 2019

Freitas, M. A. S. 1995 *Stochastische Abflussgenerieung in intermittierenden semiariden Gebieten - Nordost-Brasilien*. Hannover: Universität Hannover.

Freitas, M.A.S., 1996 *Previsão de Secas por Meio de Métodos Estatísticos e Redes Neurais e Análise de Suas Características Através de Diversos Índices (Ceará*

- *Nordeste do Brasil*), IX Congresso Brasileiro de Meteorologia, Campos do Jordão, 6 a 13 de novembro de 1996.

Freitas, M.A.S. 1997 *Regional Drought Analysis by Statistic Methods und Neuro-Fuzzy-Systems with Application to Northeast-Brazil* (in German), PhD Thesis, University Hannover, Germany.

Freitas, M. A. S., Billib, M.H.A. 1997 *Drought Prediction and Characteristic Analysis in Semi-Arid Ceará / Northeast Brazil*, Symposium "Sustainability of Water Resources Under Increasing Uncertainty", IAHS Publ. No. 240, 105-112, Rabat, Marrocos.

Freitas, M. A. S. 2005 Um Sistema de Suporte à Decisão para o Monitoramento de Secas Meteorológicas em Regiões Semi-Áridas. *Revista Tecnologia - UNIFOR*, Fortaleza, v. Suplem, p. 84-95.

Freitas, M. A. S. 2010 *Que Venha a Seca: Modelos para Gestão de Recursos Hídricos em Regiões Semiáridas*. 1. ed. Rio de Janeiro: CBJE Editora, 413p.

Freitas, M. A. S. 2016 Drought forecasting, assessment and water allocations strategies under climate change: the 2011-2015 drought period in Northeast Brazil and water crisis of the Cantareira System (São Paulo). European Conference of Tropical Ecology, Göttingen, feb.23-26, 2016.

Freitas, M. A. S. 2019 Gestão e modelagem hídrica em bacia hidrográfica do semiárido brasileiro. Brazilian Journal of Development, v.5, n.7, p.8344-8351, jul.2019.

Freitas, M. A. S., Rebello, E. R. G., Carvalho, B. E. F. C., Costa, J. A. V., Cavalcante, O. A. 2012 *Estratégias para a redução de riscos de desastres relacionados aos eventos hidrometeorológicos extremos (secas) no estado do Rio Grande do Norte Nordeste do Brasil*. In: Congresso Brasileiro Sobre Desastres Naturais, 2012, Rio Claro - São Paulo. Anais do Congresso Brasileiro Sobre Desastres Naturais. Rio Claro: Editora Unesp.

Gondim Filho, J. G. C., Sugai, M., Nobrega, M. T., Freitas, M. A. S. 2005 *Implementação e Operação do Modelo de Alocação de Água do Sistema Cantareira - São Paulo*. In: XVI Simpósio Brasileiro de Recursos Hídricos, 2005, João Pessoa. Anais do XVI Simpósio Brasileiro de Recursos Hídricos. Porto Alegre: Editora da ABRH, v. 1. p. 129-129.

Guttman, N.B. 1991 A Sensitivity Analysis of the Palmer Hydrologic Drought Index, *Water Resources Bulletin*, 27(5), 797-807.

Hastenrath, S. 1984 Interannual variability and annual cycle: mechanisms of circulation and climate in the tropical Atlantic. *Mon. Wea. Rev.*, 112, 1097-1107.

Kosko, B. 1992 *Neural Networks and Fuzzy Systems*, A Dynamical Systems Approach to Machine Intelligence, Prentice-Hall, Inc., Englewood Cliffs, N.J., USA.

Lima, T. S., Cardoso, A. S, Souza, W.M., Galvíncio, J. D. Analysis of climate variability in semiarid region, Petrolância, Pernambuco. Journal of

Hyperspectral Remote Sensing, v.6, n.2 (2016) 91-98. DOI: 10.5935/2237-2202.20160008.

Marengo, J.A. 2006 Mudanças climáticas globais e seus efeitos sobre a biodiversidade: caracterização do clima atual e definição das alterações climáticas para o território Brasileiro ao longo do século XXI, Brasília – DF.

Martorano, L. G., Vitorino, M. I., Silva, B. P. P. C., Moraes, J. R. S. C., Lisboa, l. S., Sotta, E. D. and Reichardt, K. Cimate conditions in the eastern amazon: Rainfall variability in Belem and indicative of soil water deficit. African Journal of Agricultural Research, vol. 12(21), pp.1801-1810, 25May, 2017.

Moscati, M. C. L., Gan, M. A. 2007 Rainfall variability in the rainy season of semiarid zone of Northeast Brazil (NEB) and its relation to wind regime. *International Journal of Climatology*, n. 27, p. 493-512.

Moura, A. D., Shukla, J. 1981 On the Dynamics of Droughts in Northeast Brazil: Observations, Theory and Numerical Experiments with a General Circulation Model, *Journal of the Atmospheric Sciences*, pp. 2653–2675, 1981.

Nobre, C.A., M.D. Oyama, G.S. Oliveira, J.A. Marengo, E. Salati 2004 *Impacts of climate change scenarios for 2091-2100 on the biomes of South America.* First CLIVAR International Conference, Baltimore, USA, 21-25 June.

Palmer, W.C. 1965 *Meteorological Drought*, Weather Bureau, U.S. Department of Commerce, Washington, D.C., Research Paper n° 45, 1-58.

Raha, S., Gayen, S. K. 2019 Simulation of Meteorological Drought of Bankura District, West Bengal: Comparative Study between Exponential Smoothing and Machine Learning Procedures. Journal of Geography, Environment and Earth Science International, 22(1): 1-16, 2019.

Rao, V.B, Hada, K. 1990 Characteristcs of Rainfall over Brazil Annual Variations and Connections with the Southern Oscillation. *Theoretical and Applied Climatology*, v. 42, p. 81-91.

Rasmusson, E.M., Carpenter, T.H 1983 The Relationship between Eastern Equatorial Pacific Sea Surface Temperatures and Rainfall over India and Sri Lanka, *Mon. Wea. Rev.*, 111, 517-528.

Ropelewski, C.F., Halpert, M.S. 1987 Global and Regional Scale Precipitation Patterns Associated with the El Niño / Southern Oscillation, *Monthly Weather Review*, 115, 1606-1626.

Rooy, M.P. van 1965 A Rainfall Anomaly Index Independent of Time and Space, *Notos*, 14, 43.

Salati, E., Salati, E., Campanhol, T., Villa Nova, N. 2007 *Tendências das Variações Climáticas para o Brasil no Século XX e Balanços Hídricos para Cenários Climáticos para o Século XXI.* Brasília: MMA, 186 p.

Silveira, P. B. M., Freitas, M. A. S. 2004 Políticas de Operação de Reservatório Visando Minimizar os Impactos Socio-econômicos em Situações de Escassez. In: Seminário Internacional sobre Represas y Operación de

Embalses, 2004, Puerto Iguazú. Anais do Seminário Internacional sobre Represas y Operación de Embalses. Puerto Iguazú: CACIER.

Transeau, E.N. 1905 Forest Center of Eastern America, *Amer. Naturalist*, 39, 875-889.

Uvo, C.B., Repelli, C.A., Zebiak, S.E., Kushnir, Y. The relationships between Tropical Pacific and Atlantic SST and Northeast Brazil monthly precipitation. *J. Climate*, v.11, p.551-562, 1998.

Walker, G.T. 1928 Ceará (Brazil) Famines and the General Air Movement, *Beiträge zur Physik der Atmosphäre*, 14, 88-93.

Wright, P.B. 1989 Homogeneized Long-Period Southern Oscillations Indices, *Int. Journal of Climatology*, 9, 33-54.

CHAPTER 4

The GAR(1) Model with Fragment Method for Hydrological Drought Risk Assessment in Semiarid Regions

Marcos Airton de Sousa Freitas [1], Gabriel Belmino Freitas [2]

[1]*Senior Water Resources Specialist with National Water Agency (ANA), Brazil*
[2]*Department of Economy, University of Brasília, Brazil; Master of Science student in Applied Economics at Leopold-Franzens Universität Innsbruck.*

Abstract: The Northeast Brazil has been recognized as an area that will suffer greatly from the effects of variability and climate change. This will lead to a reduction of precipitation and streamflow in the region, causing greater pressure on the scarce water resources of the region, especially on the water stored in the reservoirs. Optimization of the design and operation of multipurpose reservoir systems depends on the ability of synthetic streamflow generation models to reproduce the typical intermittent characteristics of semi-arid rivers. A Gamma Autoregressive – GAR(1) model have been tested and applied, for generating annual flows, coupled with the Fragment Method to disaggregate the annual flows to monthly ones. This coupled model was applied to four typical intermittent basins of the NE-Brazil, with drainage area varying from 410 to 5.695 Km^2. In order to analyze the performance of the model not only the statistical parameters (mean, variance, lag-1 serial correlation, etc.) of the historical and generated series were examined, but also a storage analysis by mean of the Sequent-Peak-Algorithm (SPA) was performed and additionally the preservation of the droughts and floods characteristics (duration, severity and magnitude) of the historical series was analyzed. This model was able to preserve the statistical parameters of the historical time series. However, when the generated synthetic flows were used to a storage analysis the model was not adequate to reproduce, particularly, the persistence (long periods of low and high flow) encountered in the historical series, which is fundamental by the reservoir design and for hydrological drought risk assessment.

Keywords: hydrological drought, intermittent rivers, reservoir operation optimization

1. Introduction

The Northeast Brazil has been recognized as a drought prone area (Freitas and Billib, 1997). Furthermore, this region will suffer greatly from the effects of rainfall variability and climate change. This will lead to a reduction of precipitation and streamflow in the region, causing greater pressure on the scarce water resources of the region, especially on the water stored in the reservoirs (Martins et al., 2013; Rusteberg and Freitas, 2018).

Several models are described in the literature for the generation of synthetic streamflow series, which make use of different time intervals. In general, they can be arranged into two groups (Freitas, 1995): direct simulation methods and aggregation-disaggregation methods. To the first category belong those models that generate flows simultaneously to different time intervals (Sim, 1987; Claps et al., 1993). By disaggregation models, however, the flows are originally generated for a longer time interval, e.g., one year and then they are broken down into smaller time intervals, such as month, week, day, etc. (Grygier and Stedinger, 1988; Lin and Lee, 1992).

Silva and Portela (2011) have proposed a procedure for generating synthetic series of annual and monthly flows that combines two models, a probabilistic one, applied at an annual level, and at a monthly level, a deterministic disaggregation model. The modeling of the annual flow series was based on the random sampling of the log-Pearson III law of probability. The disaggregation of annual flows into monthly flows uses the method of fragments. The combination of the two models was tested on a data set of 54 streamflow samples from gauging stations geographically spread over Mainland Portugal. For each gauging station, 1200 synthetic series were generated, with a length equal to that of the corresponding sample. The quality of the generated series was evaluated by their capacity to preserve the most significant statistical characteristics of the samples of annual and monthly flows, namely the mean, standard deviation, and skewness coefficient (Silva and Portela, 2011).

The Drought Management System (SIGES) subcomponent, referring to the analysis of hydrological droughts (Figure 1) is basically composed of the associated use of rainfall-runoff models and synthetic streamflow generation models, employed in reservoir operation simulation. These models should, however, be able to reproduce the characteristics of the drought and flood periods typical of intermittent rivers of the Brazilian semi-arid (Freitas, 2010).

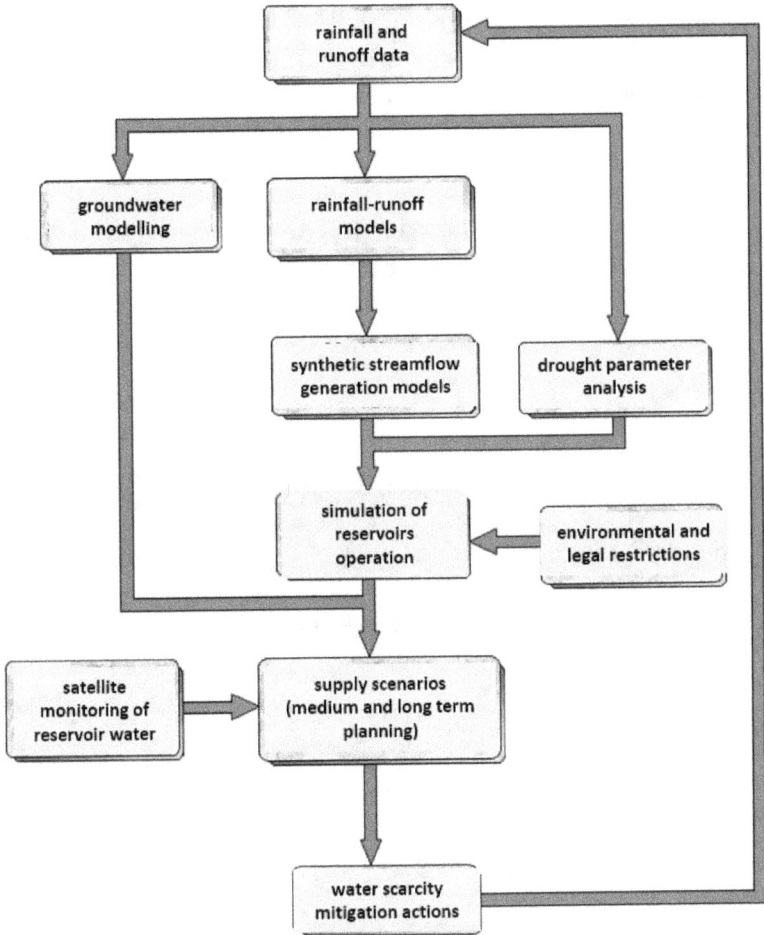

Figure 1: Schematic conception of the Drought Management System (SIGES) subcomponent concerning the hydrological drought risk assessment and reservoir operation (Freitas, 2010).

One problem faced in the streamflow droughts modeling of river belonging to semi-arid regions is the treatment of zero flows in the monthly series, due to the appearance of intermittent rivers. This paper presents a model formed by coupling an annual model (the GAR(1) model) with a monthly disaggregation streamflow model (Fragment Method).

2. GAR(1) Model

The proposed model is formed by the union of the GAR(1) - 1st order Gamma Autoregressive Model, to generate the streamflow at annual level, with the Fragments Method presented by Svanidze (1980), for disaggregation into monthly flows. A basic property of the GAR(1) model is that it does not need for transformation of the historical data flow.

According to Lawrence and Lewis (1981), a 1st order Gamma Autoregressive Model or GAR(1) model can be described by an additive process as follows:

$$Q_t = \phi Q_{t-1} + \varepsilon_t$$

$$(1)$$

with

X_t = streamflow in the time interval t;

ϕ = autocorrelation coefficient;

ε_t = independent random variable.

For the variable X_t, according to the method of moments:

$$\mu = \lambda + \frac{\beta}{\alpha}$$

$$(2)$$

$$\sigma^2 = \frac{\beta}{\alpha^2}$$

$$(3)$$

$$\lambda = \frac{2}{\sqrt{\beta}}$$

$$(4)$$

$$\rho_1 = \phi$$

$$(5)$$

where λ, α and β represent respectively the parameters of position, scale and form of a three parameters gamma probability distribution function, given by:

$$f_x(x) = \frac{\alpha^\beta (x-\lambda)^{\beta-1} \exp[-\alpha(x-\lambda)]}{\Gamma(\beta)}$$

$$(6)$$

The mean (μ), standard deviation (σ), the skewness (γ), and the lag-1 autocorrelation coefficient (ρ_1) of the population can be estimated from the sample by:

$$m = \frac{1}{N}\sum_{i=1}^{N} X_i \tag{7}$$

$$s^2 = \frac{1}{N-1}\sum_{i=1}^{N}(X_i - m)^2 \tag{8}$$

$$g_1 = \frac{N}{(N-1)(N-2)s^3}\sum_{i=1}^{N}(X_i - m)^3 \tag{9}$$

$$r_1 = \frac{1}{(n-1)s^2}\sum_{i=1}^{N-1}(X_i - m)(X_{i+1} - m) \tag{10}$$

where N is the number of years of annual flow series.

Because it is a variable with non-normal probability distribution, the values estimated by the three previous equations (7 to 9) show bias, so it is necessary to apply corrections to the values calculated above. For the autocorrelation coefficient, Wallis and O'Connell (1972) proposed the following correction:

$$\rho_1 = \frac{r_1 N + 1}{N-4} \tag{11}$$

For standard deviation, according to Matalas (1966):

$$\sigma^2 = \frac{N-1}{N-K}s^2 \tag{12}$$

where

$$K = \frac{\left[N(1-\rho_1^2) - 2\rho_1(1-\rho_1^N)\right]}{\left[N(1-\rho_1)^2\right]} \tag{13}$$

The following correction of bias in the estimation of the coefficient of skewness has been proposed by Bobée and Robitaille (1975):

$$\gamma_0 = \frac{Lg_1\left[A + B\left(\dfrac{L^2}{N}\right)g_1^2\right]}{\sqrt{N}} \tag{14}$$

where
   L = limit value of an independent asymmetric distribution;
   N = sample size.

For the gamma distribution, for values $0.25 \le \gamma \le 5$ and $20 \le N \le 90$, Bobée and Robitaille (1975) showed that the following are valid:

$$A = 1 + 6.51 N^{-1} + 20.2 N^{-2}$$ (15)

$$B = 1.48 N^{-1} + 6.77 N^{-2}$$ (16)

For values of N between 10 and 20, that due to the scarcity of streamflow measurements in intermittent rivers of northeastern Brazil is indeed common, Freitas (1995), based on an optimization procedure, proposed the following expressions to estimate the parameters A and B, for the gamma distribution:

$$A = 1 + 6.02 N^{-1} + 28.71 N^{-2}$$ (17)

$$B = 1.50 N^{-1} + 7.27 N^{-2}$$ (18)

For the lognormal distribution and values of N between 10 and 20, Freitas (1995) has proposed: $A = 0.90 + 6.70 N^{-1} + 43.00 N^{-2}$ and $B = 0.01 N^{-1} + 124.00 N^{-2}$. For the Weibull distribution and values of N between 10 and 20, Freitas (1995) has proposed the following: $A = 0.88 + 7.50 N^{-1} + 15.00 N^{-2}$ e $B = 0.30 N^{-1} + 38.00 N^{-2}$.

The skewness, however, is not only dependent on sample size N, but also on the autocorrelation coefficient. Fernandez and Salas (1990), based on Monte-Carlo simulation, have proposed a further correction to the expression given by Bobée and Robitaille (1975):

$$\gamma = \frac{\gamma_0}{f}$$ (19)

Being f the new correction factor to the following conditions $0.5 \le \gamma \le 2.0$ and $0 \le \rho_1 \le 0.8$, where:

$$f = (1 - 3.12 \rho_1^{3.7} N^{-0.49})$$ (20)

If $\rho_1 = 0$, automatically f=1, so that there is no need of correction.

For a random number generation with gamma distribution the following scheme has been used:

$$\varepsilon = \lambda(1-\phi) + \eta \qquad (21)$$

with

$$\begin{cases} \eta = 0 \rightarrow M = 0 \\ \eta = \sum_{j=1}^{M} Y_j(\phi)^{U_j} \rightarrow M > 0 \end{cases}$$

with

M = Poisson random variable with mean equal to $-\beta \ln(\phi)$

U_j = uniform random variable $(0,1)$

Y_j = random variable with exponential distribution with mean equal to $1/\alpha$.

3. Method of Fragments

The Method of Fragments from Svanidze (1980) is based on the disaggregation of annual flows generated by some annually model (in this case, the GAR(1) model) into monthly flows (or shorter time interval). The model is characterized by estimating, for each month j and each year i of the historical flow series, the so-called fragments, given by:

$$f_{i,j} = \frac{Q_{i,j}}{\sum_{j=1}^{n} Q_{i,j}} \qquad (22)$$

with:

n = number of months (n = 12); $Q_{i,j}$ = streamflow in month j of year i.

The fragments $f_{i,j}$ correspond to the percentage of annual streamflow (the denominator of the equation 22) in year i. Following the historic annual flow values are placed in ascending order and separated into classes. The limits of the class intervals are formed by the mean values of successive flows. The total number of classes is equal to the number of years within the flow series measurement. The first class has zero as the lower limit and the last class upper limit of the last class is infinite. The annual streamflows generated are then distributed according to the class intervals and fragmented into monthly values, as follows.

$$Q_{i,j} = f_j^k Q_i^k$$

(23)

with:

Q_i^k = vazão gerada pelo modelo anual, pertencente a classe k;

f_j^k = fragmento correspondente do mês j da classe k.

4. GAR(1) Model with Fragment Methods

The coupled model GAR (1) / Method of Fragments was applied to 4 semiarid basins northeastern Brazil (Sítio Poço Dantas; Faz. Cajazeiras; Sítio Novos e Limoeira), with drainage area ranging between 410 and 5.695 km² (Figure 2). To analyze the performance of the model in generating synthetic monthly streamflows the following criteria were used: i) comparing the statistical parameters (mean, standard deviation, coefficient of variation, skewness, correlation coefficient and autocorrelation coefficient) of the generated and historical series, ii) analysis of standard capabilities of reservoirs by using the Sequent-Peak-Algorithm (SPA) and iii) analysis of characteristic parameters (duration, severity and magnitude) of floods and droughts periods.

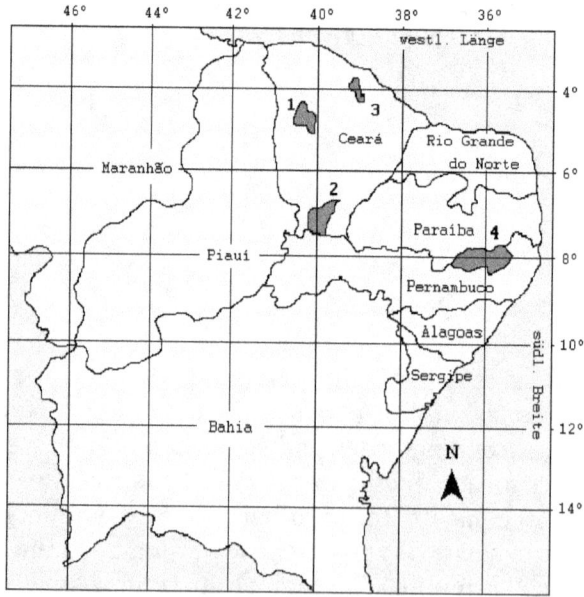

Figure 2: Location of analyzed basins.

The Sequent-Peak-Algorithm (SPA), presented by Loucks et al. (1981) and applied by Vogel and Stedinger (1987), Adeloye (1993), Freitas (1995, 2010), as well Tallaksen et al. (1996), can thus be formulated as follows:

Let Kt be the storage capacity required at the beginning of period t, Yt be the required release during the period, and Qt be the corresponding inflow during period t. If one defines:

$$K_t = \begin{cases} Y_t - Q_t + K_{t-1}; \ldots if \ldots positive \\ 0; \ldots otherwise \end{cases} for \ldots t \subseteq 2T$$

(24)

then the reservoir capacity Cp required is given by:

$$Cp = max\ (Kt);\ t = 1,2,..T,T+1,..2T$$

(25)

To implement equation 24, the reservoir is assumed to be initially full, i.e. $K_0=0.0$. Also, equation 24 is applied over 2T, where T is the length of data record, to take care of a situation in which the critical period is close to the end of record T; this wrapping round of the streamflow data record also ensures, as was noted earlier, that a steady-state solution to the storage-yield problem is obtained (Vogel, 1987; Adeloye, 1993).

Hence equation 24 is termed the double cycling SPA and the capacity Cp thus obtained represents the minimum storage required over the historic T-year period to supply the desired yield Y without shortages.

For each river basin 100 monthly series of 50-year extension have been generated. There have been calculated the minimum, maximum, median and the 25% and 75% quantile of various parameters of the series generated as well as the historical value corresponding to each month. Each of the 1000 traces (with 50 years of data extension) was routed through the reservoir using the SPA (equation 24) and the capacity required to satisfy the yield Y without failure over the T-year period was determined (equation 25).

Table 1 shows the result of 1000 yearly sets simulation, each with a 50-year extension. The values of the standardized required capacity curve (mean, standard deviation, median, 90% quantile and non-exceedance probability), determined with the series (twice as long) via Index-Sequential-Method (ISM), as Loucks et al. (1981) and the AR(1) model, according to Maass et al. (1962), but with log-gamma distribution according to Freitas (1995) and GAR(1) model, are herein presented. It is seen that the AR(1) model underestimates the standardized required capacity. Furthermore, only 15% of the generated series had more extreme dry periods that exist in the historical time series.

Table 1: Simulation of the reservoir by SPA: Capacity of the model to reproduce extreme drought periods - Cp statistics (1000 generated series), respectively, for the Sítio Poço Dantas, Faz. Cajazeiras, Sítio Novos and Limoeira stations.

Sítio Poço Dantas station

Model	Cp	Stand. deviation	median	Q90%	P(%)
ISM	4.275	0.000	4.275	4.275	0
AR(1)	4.178	2.707	3.902	6.116	45
GAR(1)	4.486	2.437	4.221	6.530	50

Faz. Cajazeiras station

Model	Cp	Stand. deviation	median	Q90%	P(%)
ISM	6.035	0.000	6.035	6.035	0
AR(1)	3.759	2.008	3.505	5.629	10
GAR(1)	4.101	2.200	3.874	6.203	15

Sítio Novos station

Model	Cp	Stand. deviation	median	Q90%	P(%)
ISM	6.718	0.000	6.718	6.718	0
AR(1)	0.880	0.095	0.827	1.279	0
GAR(1)	3.858	1.665	3.627	5.599	5

Limoeira station

Model	Cp	Stand. deviation	median	Q90%	P(%)
ISM	6.676	0.000	6.676	6.676	0
AR(1)	0.594	0.046	0.557	0.873	0
GAR(1)	3.815	1.679	3.552	5.481	5

Table 2 shows the mean statistical parameters for the duration, severity and magnitude of drought periods of the generated time series by AR(1) and GAR(1) models for Sítio Poço Dantas station, Ceará State.

Table 2: Statistical parameters of drought periods of the historical and generated series for Sítio Poço Dantas station, Ceará State.

Statistical Parameter (Drought duration)	Historical Serie	Generated Series with AR(1) model	Generated Series with GAR(1) model
mean (year)	3,667	3,323	3,041
stand. dev. (year)	1,528	2,737	2,535
variation coeff.	0,417	0,824	0,834
skewness coeff.	-0,382	1,969	2,588
lag-1 autocorrel. coeff.	-0,595	0,002	0,036

Statistical Parameter (Drought severity)	Historical Serie	Generated Series with AR(1) model	Generated Series with GAR(1) model
mean (m^3/s*year)	87,480	107,989	159,845
stand. dev. (m^3/s*year)	53,866	95,152	148,100
variation coeff.	0,616	0,881	0,927
skewness coeff.	-0,160	2,030	2,499
lag-1 autocorrel. coeff.	-0,542	0,000	0,035

Statistical Parameter (Drought magnitude)	Historical Serie	Generated Series with AR(1) model	Generated Series with GAR(1) model
mean (m^3/s)	22,173	32,290	50,496
stand. dev. (m^3/s)	6,080	11,240	18,245
variation coeff.	0,274	0,348	0,361
skewness coeff.	-0,261	-0,461	-0,631
lag-1 autocorrel. coeff.	-0,567	-0,030	0,039

Table 3 shows the mean statistical parameters for the duration, severity and magnitude of drought periods of the generated time series by AR(1) and GAR(1) models for Faz. Cajazeiras station, Ceará State.

Table 3: Statistical parameters of drought periods (duration, severity and magnitude) of the historical and generated series for Faz. Cajazeiras station, Ceará State.

Statistical Parameter (Drought duration)	Historical Serie	Generated Series with AR(1) model	Generated Series with GAR(1) model
mean (year)	4,000	3,162	3,083
stand. dev. (year)	3,000	2,912	2,587
variation coeff.	0,750	0,920	0,839
skewness coeff.	0,000	3,468	1,907
lag-1 autocorrel. coeff.	0,000	-0,003	0,009

Statistical Parameter (Drought severity)	Historical Serie	Generated Series with AR(1) model	Generated Series with GAR(1) model
mean (m^3/s*year)	263,568	183,501	258,874
stand. dev. (m^3/s*year)	175,737	191,887	250,495
variation coeff.	0,667	1,046	0,968
skewness coeff.	-0,033	3,501	1,872
lag-1 autocorrel. coeff.	0,000	-0,017	0,000

Statistical Parameter (Drought magnitude)	Historical Serie	Generated Series with AR(1) model	Generated Series with GAR(1) model
mean (m^3/s)	71,832	54,481	78,304
stand. dev. (m^3/s)	12,694	22,731	30,312
variation coeff.	0,177	0,417	0,387
skewness coeff.	0,623	-0,383	-0,539
lag-1 autocorrel. coeff.	-0,083	-0,043	-0,008

CONCLUSIONS

In general, the GAR(1) model was able to satisfactorily reproduce the statistical parameters (mean, standard deviation, etc.) the historical series, when applied to analyzed watersheds. However, as described by Freitas (1995,

2010), this single criterion is not sufficient for analyzing the performance of streamflow generation models to rivers of semiarid regions. In fact, when analyzing the results of reservoir simulation and the reproduction of the characteristics (duration, severity and magnitude) of floods and droughts periods found in the time series, the GAR(1) model can't effectively respond to these questions which it is of huge importance to the application of stochastic models in the semiarid intermittent rivers.

5. References

Adeloye, A. J. 1993 *Operational assessment of the reliability of single estimates of reservoir capacity*. Extreme Hydrological Events: Precipitation, Floods and Droughts (Proceedings of the Yokohama Symposium, July 1993). IAHS Publ. no. 213.

Bobée, B., Robitaille, R. 1975 Correction of Bias in the Estimation of the Coefficient of Skewness, *Water Resources Research*, **11**(6), 851-854.

Claps, P., Rossi, F., Vitale, C. 1993 Conceptual-Stochastic Modeling of Seasonal Runoff Using Autoregressive Moving Average Models and Different Scales of Aggregation, *Water Resources Research*, **29**(8), 2545-2559.

Fernandez, B., Salas, J.D. 1990 Gamma-Autoregressive Models for Stream-Flow Simulation, *Journal of Hydraulic Eng.*, 116, 11, 1403-1414.

Freitas, M. A. S. 1995 *Stochastische Abflussgenerierung in intermittierenden semiariden Gebieten (NO-Brasilien)*. Abschlussarbeit Weiterbildendes Studium Bauingenieurwesen, University of Hannover, Germany.

Freitas, M. A. S. 2010 Que venha a seca: modelos para gestão de recursos hídricos em regiões semiáridas (Drought Risk Analysis: Models for Water Resources Management in Semiarid Regions), 1ª ed., Rio de Janeiro, 413p.

Freitas, M. A. S., Billib, M.H.A. 1997 *Drought Prediction and Characteristic in Semi-Arid Ceará / Northeast Brazil*, Symposium "Sustainability of Water Resources Under Increasing Uncertainty", IAHS Publ. nº 240, 105-112, Rabat, Marrocos.

Grygier, J. C., Stedinger, J. R. 1988 Condensed Disaggregation Procedures and Conservation Corrections for Stochastic Hydrology, *Water Resources Research*, **24**(10), 1574-1584.

Lawrence, A. J., Lewis, P.A.W. 1981 A New Autoregressive Time Series Model in Exponential Variables NEAR(1), *Adv. Appl. Prob.*, **13**(4), 826-845.

Lin, G-F., Lee, F-G. 1992 Multistage Disaggregation Process in Stochastic Hydrology, *Water Resources Management*, 6, 101-115.

Loucks, D. P., Stedinger, J. R., Haith, D. A. 1981 *Water Resource Systems Planning and Analysis*, Prentice Hall Inc., Englewood Cliffs, New Jersey.

Maass, A., Hufschmidt, M. M., Dorfman, R., Thomas, H. A., Marglin, S. A. & Fair, G. M. 1962 *Design of Water Resource Systems*, Harvard University Press, Cambridge, Massachusetts.

Martins, E.S.; Braga, C.F.C; Souza, F.A.S.; Moraes, M.M.G.A.; Marques, G.F.; Mediondo, E.M.; Freitas, M.A.S.; Vazquez, V.; Engle, N.L.; De Nys, E. (2013). Adaptation challenges and opportunities in Northeast Brazil. Environment and Water Resources Occasional Paper Series, v. 11, n. 1, p. 1-6, 2013.

Matalas, N.C. 1966 Time Series Analysis, *Water Resources Research*, **3**(4), 817-829.

Rusteberg, B. and Freitas, M. A. S. 2018. *IWRM Implementation in North-East Brazil (Results from WP 8)*. In: Abels, A., Freitas, M. A. S., Pinnekamp, J., Rusteberg, B. *Bramar Project – Water Scarcity Mitigation in Northeast Brazil*, 1. ed. Aachen: Department of Environmental Engineering (ISA), 2018. v. 1. 155p. ISBN: 978-3-00-059926-2.

Silva, A. T., Portela, M. M. 2011. Generation of monthly synthetic streamflow series based on the method of fragments. *WIT Transactions on Ecology and the Environment*, Vol 145, 237-247. doi:10.2495/WRM110201.

Sim, C. H. 1987 A Mixed Gamma ARMA(1) Model for River Flow Time Series, *Water Resources Research*, **23**(1), 32-36.

Svanidze, G.G. 1980 *Mathematical Modeling of Hydrologic Series for Hydroelectric and Water Resources Computations*, Water Resources Publications, Fort Collins, Colorado, USA.

Tallaksen, L.M., Madsen, H., Clausen, B. 1996 On the definition and modelling of streamflow drought duration and deficit volume, *Hydrological Sciences-Journal-des Sciences Hydrologiques*, **42**(1), 15-33.

Vogel, R. M. 1987 Reliability indices for water supply systems, *J. Wat. Resour. Plan. Manage.*, ASCE, 113, 563-579.

Vogel, R. M., Stedinger, J. R. 1987 Generalized storage-reliability-yield relationships. *Journal of Hydrology*, 89, 303-327.

Wallis, J. R., O'Connell, P.E. 1972 Small Sample Estimation of ρ_1, *Water Resources Research*, 8(3).

CHAPTER 5

On the Applicability of Multi-seasonal Streamflow Generation Models for Intermittent Rivers

Marcos Airton de Sousa Freitas [1], Gabriel Belmino Freitas [2]

[1]*Senior Water Resources Specialist with National Water Agency (ANA), Brazil*
[2]*Department of Economy, University of Brasília, Brazil; Master of Science student in Applied Economics at Leopold-Franzens Universität Innsbruck.*

Abstract: *Monthly and annually streamflow generation models have been used for Monte Carlo simulations, in order to design, optimization and risk analysis of multipurpose reservoirs systems. Auto-regressive models, in general, preserve the statistical parameters of the historical time series when they are applied in humid watersheds. However, they are not able to reproduce the persistence encountered in the historical series of intermittent rivers from semi-arid areas. In this study is presented one application of several streamflow generation models for semi-arid area of the Northeast Brazil. Frag1 model and Frag2 model showed the best results. The Matalas model also proved to be quite efficient in reproducing streamflow. For water resources management in semi-arid regions, hydrologic droughts risk assessment is of huge importance for water resources policy makers.*
Keywords: *water resources, optimization, semi-arid regions, streamflow generation*

1. Introduction

The brazilian semi-arid region has an area of about 1 million km². It is characterized by strong temporal and spatial rainfall variability (400 to 1800 mm/year) and a high evaporation rate (above 2000 mm/year), associated with geological restrictive conditions (crystalline basement of reduced hydrological potential), that causes river intermittency. The construction of artificial dams along the major rivers of the region throughout the last century, was indispensable for water supply, especially during drought periods.

Droughts in Northeast Brazil, which tend to intensify due to climate change (Martins et al., 2013), have repeatedly brought famine, mass migration and social conflicts in this region. Its prediction, monitoring and management, however, remain a central research theme. In water resources management in semiarid regions such as the Northeast of Brazil, it is fundamental to have tools to aid decision making (Freitas, 2010).

Several studies have indicated the influence of numerous atmospheric phenomena on rainfall in Northeast Brazil (Hastenrath, 1984; Freitas and Billib, 1997; Andreoli and Kayano, 2007; Moscati and Gan, 2007; Rusteberg and Freitas, 2018). Climatological studies have indicated the existence of a strong relationship between sea surface temperature distribution (SST - sea surface temperature) along the tropical Atlantic basin temperature and the semiarid northeastern Brazil precipitation, as well as a decadal trend associated with changes in the meridional position of the ITCZ - Intertropical Convergence Zone (Moura and Schukla, 1981; Rao and Hada, 1990; Billib and Freitas, 1996). These phenomena are indicative to be related to climate variability and extreme droughts and floods in the region.

For the design and operation of multi-purpose and multi-use surface reservoirs systems (water supply, irrigation, energy production, etc.) deterministic rainfall-runoff models and stochastic streamflow generation models are usually employed, depending on data availability (Freitas, 1996).

For generating synthetic streamflow several models with different time intervals have been reported in the literature. In general, these models can be grouped into two categories: direct generation models and disaggregation's models. In the first class belong models, which generate flow simultaneously for different time intervals (Fernandez and Salas, 1986; Sum, 1987; Bartolini et al., 1988; Claps et al., 1993). By disaggregation's models discharges are generated initially for a longer time period, e.g. a year, and then broken into smaller time intervals, such as monthly, weekly, daily, etc.

According to Dracup et al. (1980) four basic considerations must be evaluated for the definition of droughts, that are: 1) what is the greatest interest in the analysis, i.e. what is the nature of the deficit of water to be investigated (meteorological, hydrological or agricultural); 2) what is the time series discretization used in the analysis (annual, semiannual, monthly, etc.); 3) what is the threshold level of separation between flood and drought events; and 4) the choice of the regionalization and standardization methods to be adopted.

A hydrological drought can be defined as a one or more sequenced years, when the average annual flow remains below the long term mean annual flow, considering all the existing series (Dracup et al., 1980). A drought event may thus be characterized by three parameters, namely the duration D in years; the accumulated deficit or severity S and the magnitude M, which represents the average cumulative deficit below the mean annual flow.

When applying stochastic models for generating streamflow time series it is necessary to observe not only the characteristics of the streamflow time series, but also the use for which they are intended. One of the most important aspects in the analysis of water resources in semiarid regions are the impacts of extreme events, in particular prolonged droughts, on water resources systems.

For this it is essential to generate long time series of synthetic streamflows. Askew et al. (1971), Stedinger and Taylor (1982), as well as Dracup and Kendall (1992) discussed the inability of traditional models, based on Markov chain, to reproduce the frequency distribution of extreme drought events that occurred in the historical series.

The SAGE software consists of models (Figure 1), adapted from models mentioned in the literature to reproduce the typical characteristics of intermittent rivers in the semiarid region. The software, in its current version, was written in Visual Basic language, in such a way that it, through a user-friendly interface, could be used as a support tool in watershed committees in the generation of synthetic flow series necessary for the simulation of reservoir operation on grant (authorization) studies, among others.

2. Streamflow Generation Models

The SAGE software consists of models (Figure 1), adapted from models cited in the literature to reproduce the typical characteristics of intermittent rivers of the semiarid regions. The software, in its current version, was written in Visual Basic language, in such a way that it, through a user-friendly interface, could be used as a support tool in watershed committees in the generation of synthetic flow series necessary for the simulation of reservoir operation on grant (authorization) studies, among others.

A first annual model to be described is called Thomas-Fiering model or AR(1) model, i.e. Auto-Regressive of order 1, which is based on a stochastic process (Mass et al., 1962). The second model is the Gamma Autoregressive or GAR(1) model, proposed by Fernandez and Salas (1990). Both models are special cases of the ARMA model (Box and Jenkins, 1976).

2.1   Annual Model

2.1.1 AR(1) Model

The original AR(1) model, also known as Thomas-Fiering Model, can be described by the following equation:

$$Q_i = \mu + \rho(Q_{i-1} - \mu) + t_i \sigma (1 - \rho^2)^{1/2} \qquad (1)$$

with

Q_i = streamflow in year i;
μ = population mean;

σ = standard deviation of the population;

ρ = lag-one correlation coefficient of the population;

t_i = random variable N(0,1).

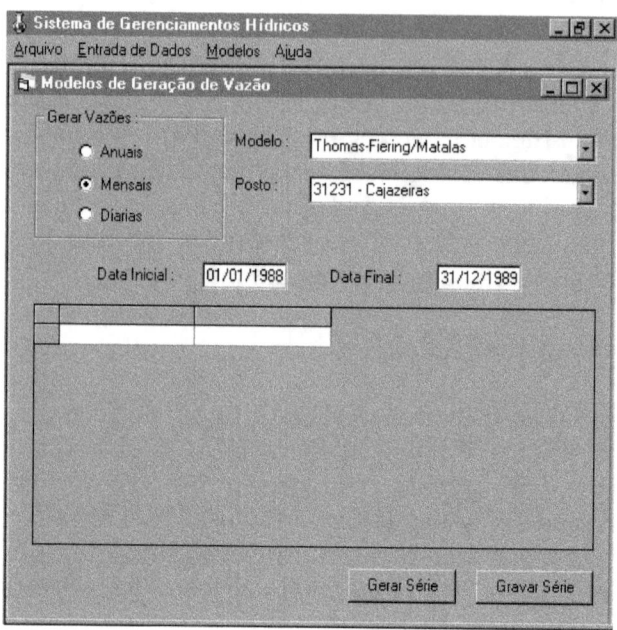

Figure 1: The SAGE Software for Streamflow Generation

A first annual model to be described is called Thomas-Fiering model or AR(1) model, i.e. Auto-Regressive of order 1, which is based on a stochastic process (Mass et al., 1962). The second model is the Gamma Autoregressive or GAR(1) model, proposed by Fernandez and Salas (1990). Both models are special cases of the ARMA model (Box and Jenkins, 1976).

2.1.2 GAR(1) Model

A 1st order Gamma Autoregressive Model or GAR(1) model can be described by an additive process as follows:

$$Q_i = \phi Q_{i-1} + \varepsilon_i \tag{2}$$

with

Q_i = streamflow in the time interval i;

ϕ = autocorrelation coefficient;

ε_i = independent random variable.

For random number generation with gamma distribution the following scheme has been used:

$$\varepsilon = \lambda(1-\phi) + \eta \qquad (3)$$

where

$$\begin{cases} \eta = 0 \rightarrow M = 0 \\ \eta = \sum_{j=1}^{M} Y_j(\phi)^{U_j} \rightarrow M > 0 \end{cases}$$

with

M = Poisson random variable with mean equal to $-\beta \ln(\phi)$

U_j = uniform random variable (0,1)

Y_j = random variable with exponential distribution with mean equal to $1/\alpha$.

2.1.3 ARR (Alternating Renewal Reward) Model

The annual model Alternating Renewal Reward by Dracup and Kendall (1992) is based on the characteristics (duration, severity and magnitude) of droughts and floods periods found in the historical series. It makes use of the geometric distribution to simulate the droughts and floods duration and the gamma distribution with two parameters for reproduction of the severity.

A basic assumption on modeling process of annual flows through the ARR model is that drought events come from different populations, ie, the deficit Yi (deficit in year i) is uniformly distributed and independent, dependent, however, on duration. For the annual flow generation two stages are performed: 1) drought flood modeling process, 2) flow modeling within drought or flood periods. The model can therefore be found in one of two possible stages. If, for example, the system is assumed a priori to be flood, then DH_1 years of flood are generated. The following step adopted is to assume the system is to be in drought condition, and DL_1 years of drought are generated, and so forth. DH_n and DL_n are intended to have independent and uniformly distributed.

Thus, the problem consists only in identifying the probability distribution functions for the two stages model. For the flood/drought modeling process, geometric distributions have been used and for the modeling of flood and drought severity, two-parameter gamma distribution have been applied. A limiting factor for the adjusting of duration (length) and severity distributions is the small sample. During a about 80 years data period, for example, there are approximately only 15 drought periods. To overcome this deficiency two procedures have been employed: decimation and standardization. In the analysis of time series with time interval less than one year, that is, if there is periodicity, the original variable can be replaced by a standardization procedure to remove this periodicity.

The decimation procedure (Bloomfield, 1976) was used to obtain n (number of months) streamflow series, through the use of standardized flow values of the straight years, for each drought (flood), to each of the n series. The drought and flood periods severities are thus evaluated. This procedure simulates a regionalization through n flows sets of n different positions of a homogeneous region, subject to the same climatic conditions.

To simulate flood and drought duration geometric distributions was used, as follows:

$$f_x(x) = pq^{x-1}$$

(4)

For the severity two-parameter gamma distributions for each duration (time length), was employed, as Kendall and Dracup (1992):

$$f(Y_D) = \frac{\lambda e^{-\lambda Y_D}(\lambda Y_D)^{r-1}}{\Gamma(r)}$$

(5)

with

Γ = gamma function; r = shape parameter; λ = scale parameter.

2.2 Monthly Model

2.2.1 PAR(1) Model

The monthly Thomas-Fiering model or PAR (1) can be represented by the following equation:

$$Q_{i,j} = \overline{Q} + \frac{s_j}{s_{j-1}} r_j (Q_{i,j-1} - \overline{Q}_{j-1}) + t'_j s_j (1 - r_j^2)^{1/2}$$

(6)

with

$Q_{i,j-1}$ e $Q_{i,j}$ = flow in month j-1 and j of year i

$\overline{Q}_{j-1}$ e $\overline{Q}_j$ = average flow of months j-1 and j

s_{j-1} e s_j = standard deviation of months j-1 e j

r_j = correlation coefficient of the month j

t_j = random variable of an asymmetric distribution

Unsatisfactory results arising from applications of the conventional Thomas-Fiering model in semi-arid regions, with many zero flows, brought proposals for modifications of this model presented by Clarke (1973) and Filho (1978). These procedures take into accounting the independence between the occurrence or non-occurrence of flow in each month of the year.

Initially probabilities of occurrences of flow for each month have been determined, given by:

$$P_j = \frac{n'}{n}$$

(7)

with

n′ = number of non-zero values of flow in a given month j with n observed data;

n = number of years in the historical time series

For each month a random number with uniform distribution Vj (0,1) is then generated and compared to a value of probability of flow.

$$V_j > P_j \Rightarrow Q_{i,j} = 0$$
$$V_j \leq P_j \Rightarrow Q_{i,j} > 0$$

when $Q_{i,j} > 0$ e $Q_{i,j-1} = 0$, then $Q_{i,j}$ is determined as follows:

$$Q_{i,j} = \overline{Q}_j + s_j t_j$$

(8)

with

t_j = random number with a normal distribution N (0,1).

Another modification of this model was proposed by Matalas (1967). A lognormal probability distribution is often used. Not only due to characteristics of asymmetric flows monthly, but also due to difficulty of obtaining a representative value estimator population skewness, since the flow time series are usually short. As the logarithm of zero is not defined, which would make it impossible to apply directly the log transformation to historical series of intermittent rivers, Matalas (1967) presented a way to estimate the parameters of the series in the logarithmic domain without changing the series itself.

2.2.2 Two-tier Model

However, most models do not capture the distribution and persistence of the annual flow. The monthly generated flows (by monthly models) when they are combined, normally differ from synthetic annual flows (generated by annually models), particularly when the model is specified in terms of logarithms of streamflow, or some other transformation effected. In such cases, either the monthly or annual flows need to be adjusted to maintain such consistency. A relevant question is how to adjust the seasonal flows generated without substantially distorting their marginal distributions. In this study various adjustment methods have been tested and analyzed.

By the application of stochastic flow models in water resources systems is important, therefore, that not only the statistical parameters of monthly flows, but also the annual flows, are reproduced. In general, negative flows have been generated and the distortions in the marginal distribution have been also observed. The use of an adjustment procedure is of paramount importance, especially in evaluating the ability of the generation model to reproduce the extreme drought periods and estimate with reasonable accuracy the vicinity of historical annual flows.

2.2.3 Method of Fragments

The Method of Fragments from Svanidze (1980) is based on the disaggregation of annual flows generated by some annually model (in this case, the ARR model) into monthly flows (or shorter time interval). The model is characterized by estimating, for each month j and each year i of the historical flow series, the so-called fragments, given by:

$$f_{i,j} = \frac{Q_{i,j}}{\sum_{j=1}^{n} Q_{i,j}}$$

(9)

where:

n = number of months (n = 12); $Q_{i,j}$ = flow in month j of year i.

The fragments $f_{i,j}$ correspond to the percentage of annual streamflow (the denominator of the equation above) in year i. Following the historic annual flow values are placed in ascending order and separated into classes. The limits of the class intervals are formed by the mean values of successive flows. The total number of classes is equal to the number of years within the flow series measurement. The first class has zero as the lower limit and the last class upper limit of the last class is infinite. The annual flows generated are then distributed according to the class intervals and fragmented into monthly values.

2.2.4 The Disaggregation Model

The disaggregation model proposed by Valencia and Schaake (1973) uses an annual flow generation model, and then disaggregate these annual streamflow values into monthly, weekly or daily streamflow values. This model is based on the following equation:

$$Q_i = AM_i + BV_i \tag{10}$$

with

Q_i = matrix flow
A = parameter vector [12x1]
M_i = column matrix flow in year i subtracted from μ estimated by

$$\overline{M} = \frac{1}{12} \sum_{j=1}^{12} \overline{Q}_j$$

B = parameter matriz [12x12]
V_j = random component vector

The parameter vector A is estimated by $A = E(Q_iM_i)E^{-1}(M_iM_i)$, where E( ) is the expected value and E-1( ) is the inverse matrix of the expected value. The parameter array B is in turn determined from the following expression, which can be obtained by spectral decomposition or principal component analysis:

$$BB^T = E(Q_iQ^T_i) - E^{-1}(M_iM_i)E(Q_iM_i)E^T(Q_iM_i) \tag{11}$$

3. Streamflow Generation Models

The various models have been consolidated into a computer package called SAGE (<u>S</u>tochastische <u>A</u>bfluss<u>GE</u>nerierungsmodell), composed of the following synthetic flow generation models (Freitas, 1995; Freitas, 2010):

- PAR-model (Thomas/Fiering) with modification by Clarke (1973)
- PAR-model (Thomas/Fiering) with modification by Matalas (1967)
- Two-tier model (PAR(1)/AR(1) with log-gamma distribution)
- Two-tier model (PAR(1)/AR(1) with log-normal distribution)
- Two-tier model - PAR(1)/GAR(1) by Fernandez and Salas
- Fragment method for AR(1) with log-gamma distribution
- Fragment method for AR(1) with log-normal distribution
- Fragment method for GAR(1)
- Disaggregation model / AR(1) by Valencia and Schaake (1973)
- ARRF model (Alternating Reward Renewal Model / Fragment method)

In order to verify the applicability of these models to the Brazilian semiarid intermittent rivers, they have been applied to four representative northeastern Brazil rivers (Figure 2), these basins with areas ranging from 410 to 5695 km^2 (Table 1).

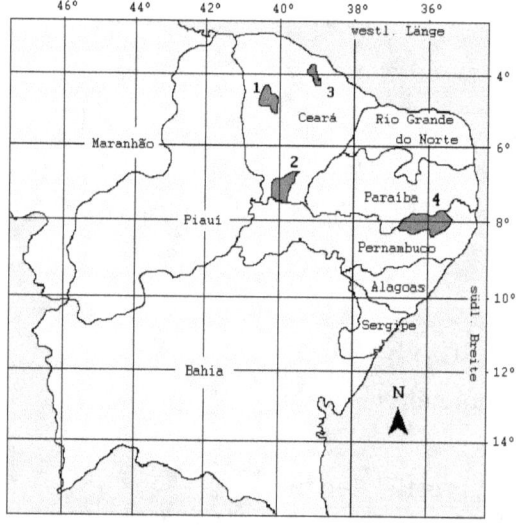

Figure 2: *Location of analyzed basins.*

To analyze the performance of flow generation models in semiarid regions three basic criteria are needed, namely: (1) analysis of the statistical parameters of the generated series, (2) analysis of the result of the reservoir operation simulation and (3) analysis of the characteristics (duration, severity and magnitude) of drought and flood generated periods (Freitas, 1995).

Table 1: Characteristics of analyzed hydraulic basins

Nr.	Station	River	Basin area (km2)	Annual average flow (m3/s)	Period
1	Faz. Cajazeiras	Acaraú	1550	7.45	1963-82
2	Sitio Poço Dantas	Bastiões	3700	3.88	1968-81
3	Sitio Novos	São Gonçalo	410	3.05	1963-75
4	Limeira	Capibaribe	5695	6.55	1957-75

For each basin were generated 100 series with 50-years of extension, by each model. In Figure 3 are shown the historical value, the median value (100 series generated), mean and standard deviation, respectively. The Frag1 and Frag2 models were that best reproduced the analyzed statistical parameters.

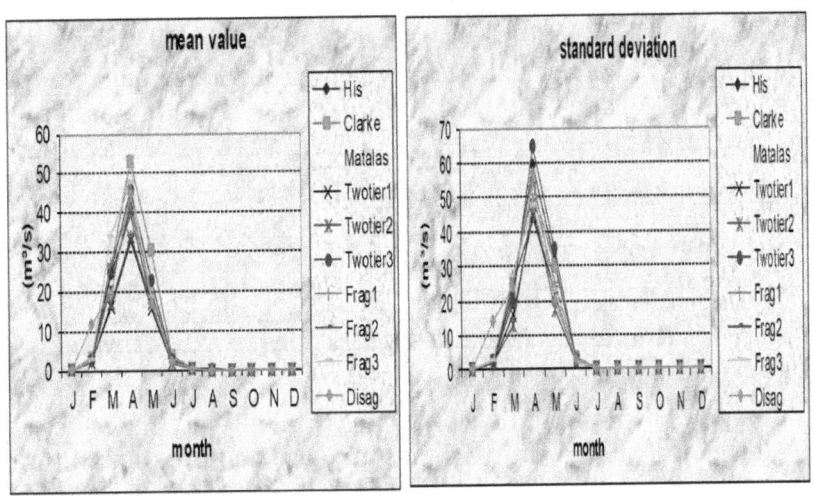

Figure 3: Statistical parameters of the generated and historical series (mean value and standard deviation)

Table 2 shows the characteristics of historic and generated series for the Faz. Cajazeiras station, on Acaraú river. In Tables 3-5 are the statistics of the characteristic parameters of hydrological droughts and floods periods (duration, severity and magnitude) for annually models.

Table 2: Comparisons of the statistical characteristics of the historical and generated series for Faz. Cajazeiras station.

Statistical paremeters	historic serie	generated serie PAR(1)	GAR(1)
mean (m^3/s)	91.384	98.882	115.720
standard deviation (m^3/s)	104.223	96.545	122.283
variation coefficient	1.140	0.976	1.057
asymmetry coefficient	1.579	1.923	1.607
lag-1 correlation coefficient	0.265	0.244	0.270

Table 3: Comparison of the statistics of the hydrological drought duration of the historical and generated series for Faz. Cajazeiras station.

Statistical paremeters	historic serie	generated serie PAR(1)	GAR(1)
mean (year)	3.133	3.053	3.141
standard deviation (year)	2.232	2.445	2.603
variation coefficient	0.712	0.801	0.829
asymmetry coefficient	1.184	1.692	1.909
lag-1 correlation coefficient	0.451	-0.004	0.018

Table 4: Comparison of the statistics of the hydrological drought severity of the historical and generated series for Faz. Cajazeiras station.

Statistical paremeters	historic serie	generated serie PAR(1)	GAR(1)
mean (m^3/s*year)	204.761	181.444	244.755
stand. deviation (m^3/s*year)	172.733	171.353	239.075
variation coefficient	0.844	0.944	0.977
asymmetry coefficient	1.231	1.714	1.935
lag-1 correlation coefficient	0.393	-0.025	0.011

Table 5: Comparison of the statistics of the hydrological drought magnitude of the historical and generated series for Faz. Cajazeiras station.

Statistical paremeters	historic serie	generated serie PAR(1)	GAR(1)

mean (m³/s)	59.581	55.336	71.772
standard deviation (m³/s)	22.164	24.068	29.513
variation coefficient	0.372	0.435	0.411
asymmetry coefficient	-0.839	-0.252	-0.494
lag-1 correlation coefficient	0.183	-0.033	-0.058

Table 6 shows the average adjustment errors (bias and rmse) for Faz. Cajazeiras station.

Table 6: Average adjustment errors (bias and rmse) for Faz. Cajazeiras station

model	mean		standard deviation		lag-1 correlation coef.	
	bias	rmse	bias	rmse	bias	rmse
TF/Clarke	0.2876	0.2321	0.2060	0.1674	6.6039	4.1213
TF/Matalas	0.1965	0.1953	0.1639	0.1591	6.4141	1.7117
Two-tier1	0.3237	0.2652	0.2810	0.2007	7.1211	3.9973
Two-tier2	0.3031	0.2808	0.2369	0.2922	8.7538	2.2963
Two-tier3	0.3325	0.2704	0.2691	0.3102	8.5416	5.7129
Frag1	0.0966	0.2136	0.0838	0.1673	0.1276	1.9774
Frag2	0.1381	0.1825	0.1387	0.1028	0.8318	2.6558
Frag3	0.2076	0.1801	0.1111	0.1304	1.1984	1.9313
Disag.	0.4867	0.4691	0.3299	0.3379	7.3794	7.1152

4. Conclusions

In this work one application of the SAGE software was presented, as well as the description of the models used to flow generate in semiarid region. When applied to the various basins of the semi-arid the Frag1 and Frag2 models showed better results, thereby proving to be a useful tool in the design and optimization of reservoir systems in semi-arid regions. The Matalas model also proved to be quite efficient in reproducing flow in semi-arid northeastern Brazil. Freitas (2016) and Rusteberg and Freitas (2018) applied the SAGE system successfully.

5. Acknowledgements

Authors wish to thank the engineer Paulo Breno de Moraes Silveira by relevant comments and suggestions.

6. References

[1] Andreoli, R. V., Kayano, M. T. 2007 A Importância relativa do Atlântico Tropical Sul e Pacífico Leste na Variabilidade de Precipitação do Nordeste do Brasil (The relative importance of the tropical South atlantic and

eastern pacifico on rainfall variability in the northeast Brazil), *Revista Brasileira de Meteorologia*, v.22,n.1,63-74.

[2] Askew, A.J., Yeh, W.W-G., Hall, W.A. 1971 A Comparative Study of Critical Drought Simulation, *Water Resources Research, 7(1), 52-62.*

[3] Bartolini, P., Salas, J.D., Obeysekera, J.T.B. 1988 Multivariate Periodic ARMA(1,1) Processes, *Water Resources Research, 24(8), 1237-1246.*

[4] Billib, M. H. A., Freitas, M. A. S. 1996 Drought Forecasting and Management for Northeast-Brazil by Statistics, Neuro-fuzzy Systems Analysis and Stochastic Simulation. In: *Conference on Water Resources & Environment Research: towards the 21st Century*, 1996, Kyoto.

[5] Bloomfield, P. 1976 Fourier Analysis of Time Series: An Introduction, Wiley-Interscience, 258p.

[6] Askew, A.J., Yeh, W.W-G., Hall, W.A. 1971 A Comparative Study of Critical Drought Simulation, *Water Resources Research, 7(1), 52-62.*

[7] Box, G.E.P, Jenkins, G.M. 1976 Time Series Analysis: Forecasting and Control, Holden-Day Verlag.

[8] Claps, P., Rossi, F., Vitale, C. 1993 Conceptual-Stochastic Modeling of Seasonal Runoff Using Autoregressive Moving Average Models and Different Scales of Aggregation, *Water Resources Research, 29(8), 2545-2559.*

[9] Clarke, R.T. 1973 Mathematical Models in Hydrology, FAO Irrigation and Drainage Paper, n. 19.

[10] Dracup, J. A., Lee, K. S., Paulson, E.G. 1980 On the Definition of Droughts, *Water Resources Research, 16(2), 297-302.*

[11] Fernandez, B., Salas, J. D. 1986 Periodic Gamma Autoregressive Processes for Operational Hydrology, *Water Resources Research, 22(10), 1385-1396.*

[12] Fernandez, B., Salas, J. D. 1990 Gamma-Autoregressive Models for Stream-Flow Simulation, *Journal of Hydraulic Engineering, 116(11), 1403-1414.*

[13] Filho, A.A.C. 1978 Um Modelo de Simulação e Operação de um Sistema de Irrigação com Reservatórios em Rios Intermitentes (A Simulation and Operation Model of an Intermittent Reservoir Irrigation System), Master's Dissertation, UFRJ, Rio de Janeiro.

[14] Freitas, M. A. S. 1995 Stochastische Abflussgenerierung in intermittierenden semiariden Gebieten (NO-Brasilien). Abschlussarbeit Weiterbildendes Studium Bauingenieurwesen, Univ. Hannover, Deutschland.

[15] Freitas, M.A.S. 1998 A Decision Support System for Drought Forecasting and Reservoirs Management in Northeast Brazil, *VIII Congresso Latino-Americano e Ibérico de Meteorologia, X Congresso Brasileiro de Meteorologia, Brasilia - DF.*

[16] Freitas, M. A. S. 2010 *Que venha a seca: modelos para gestão de recursos hídricos em regiões semiáridas (Drought Risk Analysis: Models for Water Resources Management in Semiarid Regions)*, 1ª ed., Rio de Janeiro, 413p.

[17] Freitas, M. A. S. 2016 Drought forecasting, assessment and water allocations strategies under climate change: the 2011-2015 drought period in Northeast Brazil and water crisis of the Cantareira System (São Paulo). European Conference of Tropical Ecology, Göttingen, feb.23-26, 2016.

[18] Freitas, M. A. S., Billib, M.H.A. 1997 *Drought Prediction and Characteristic in Semi-Arid Ceará / Northeast Brazil*, Symposium "Sustainability of Water Resources Under Increasing Uncertainty", IAHS Publ. nº 240, 105-112, Rabat, Marrocos.

[19] Hastenrath, S. 1984 Interannual variability and annual cycle: mechanisms of circulation and climate in the tropical Atlantic., *Mon. Wea. Rev.*, 112, 1097-1107.

[20] Kendall, D.R., Dracup, J. A. 1992 On the Generation of Drought Events Using an Alternating Renewal-Reward Model, *Stochastic Hydrol. Hydraul.*, 6, 55-68.

[21] Lee, K.S., Dracup, J.A. 1982 A Stochastic Frequency Analysis of Multiyear Droughts, National Science Foundation, University of California Water Resources Center.

[22] Lee, K.S., Sadeghipour, J., Dracup, J.A. 1986 An Approach for Frequency Analysis of Multiyear Drought Durations, *Water Resources Research, 22(5)*, 655-662.

[23] Martins, E. S. P. R., Braga, C. F. C., De Nys, E., Souza Filho, F. A., Freitas, M. A. S. 2013 Impacto das Mudanças do Clima e Projeções de Demanda Sobre o Processo de Alocação de Água em Duas Bacias do Nordeste Semiárido (Impact of Climate Change and Demand Projections on the Water Allocation Process in Two Semi-Arid Northeast Basins) – 1ª Ed. – Brasília – 2013 112p. ISBN 978-85-6387.

[24] Mass, A. et al., 1962 Design of Water Resources Systems, Harvard University Press.

[25] Matalas, N.C. 1967 Mathematical Assessment of Synthetic Hydrology, *Water Resources Research, 3(4)*, 937-945.

[26] Moscati, M. C. L., Gan, M. A. 2007 Rainfall variability in the rainy season of semiarid zone of Northeast Brazil (NEB) and its relation to wind regime. *International Journal of Climatology, n. 27, p. 493-512.*

[27] Moura, A. D., Shukla, J. 1981 On the Dynamics of Droughts in Northeast Brazil: Observations, Theory and Numerical Experiments with a General Circulation Model, *Journal of the Atmospheric Sciences*, pp. 2653–2675, 1981.

[28] Rao, V.B, Hada, K. 1990 Characteristcs of Rainfall over Brazil Annual Variations and Connections with the Southern Oscillation. *Theoretical and Applied Climatology*, v. 42, p. 81-91.

[29] Rusteberg, B. and Freitas, M. A. S. 2018. *IWRM Implementation in North-East Brazil (Results from WP 8)*. In: Abels, A., Freitas, M. A. S., Pinnekamp, J., Rusteberg, B. *Bramar Project – Water Scarcity Mitigation in Northeast Brazil*, 1. ed. Aachen: Department of Environmental Engineering (ISA), 2018. v. 1. 155p. ISBN: 978-3- 00-059926-2.

[30] Sim, C. H. 1987 A Mixed Gamma ARMA(1) Model for River Flow Time Series, *Water Resources Research, 23(1), 32-36.*

[31] Stedinger, J. R., Taylor, M. R. 1982 Synthetic Streamflow Generation, 1: Model Verification and Validation, *Water Resources Research, 18(4), 909-918.*

[32] Svanidze, G.G. 1980 Mathematical Modeling of Hydrologic Series for Hydroelectric and Water Resources Computations, Water Resources Publications, Fort Collins, Colorado, USA.

[33] Valencia, D., Schaake, J.C. 1973 Disaggregation Processes in Stochastic Hydrology, *Water Resources Research, 9(3), 580-585.*

CHAPTER 6

Hydrological Drought Assessment: The Use of the ARRF Model for Monthly Streamflow Generation on Intermittent Rivers of the Northeast Brazil

Marcos Airton de Sousa Freitas[1], Gabriel Belmino Freitas[2]

[1](Senior Water Resources Specialist with National Water Agency)
[2](Department of Economy, University of Brasília - UnB)
Corresponding Author: National Water Agency – Adress code: 70.610-200; masfreitas@ana.gov.br

ABSTRACT: This paper presents the ARR model (Alternating Renewal Reward) for generating annual streamflow, coupled with the Fragment model to disaggregate the annual streamflow to monthly ones. This new coupled model, namely ARRF model, has been applied to three typical intermittent basins in the State of Ceará - Brazil. To better understand the phenomenology of the processes involved in modeling, several statistical tests were applied to flood and drought events of the typical hydrological streamflow series of this region. The ARRF model was able to preserve the analyzed statistical parameters of the historical streamflow series, as well as to reproduce the persistence (long periods of low and high flow, i.e. drought and flood periods) encountered in the historical flow series, which are fundamental for hydrological drought assessment and for the design and operation optimization of multi-purpose reservoirs system.

KEYWORDS: Streamflow Generation and Simulation, Semiarid Regions, Reservoir Operation Optimization

1. Introduction

The Brazilian semiarid region has an area of about 1 million km^2. It is characterized by strong temporal and spatial rainfall variability (400 to 1800 mm/year) and a high evaporation rate (above 2000 mm/year), associated with geological restrictive conditions (crystalline basement of reduced hydrological potential), that causes river intermittency. The construction of artificial dams along the major rivers of the region throughout the last century, was indispensable for water supply, especially during drought periods. Since 2011, this entire region has been suffering from a severe drought event [1]. In addition, [2], [3] and [4] demonstrated that this region will suffer even more due to climate change.

For hydrological drought assessment and for the design and operation of multi-purpose and multi-use surface reservoirs systems (water supply, irrigation, energy production, etc.) deterministic rainfall-runoff models and stochastic streamflow generation models are usually employed [5], depending on data availability, which is generally scarce in this region.

Several autoregressive models have been presented in the literature for generating synthetic streamflow in temperate regions. For semiarid regions, however, such models cannot reproduce satisfactorily the typical characteristics of intermittency. For this purpose, [6] has modified and applied nine autoregressive models to monthly flow generation in intermittent rivers of four basins in the semiarid region of Brazil. The analyzed models showed, in most cases, an overestimate of the average monthly flow, probably due to their Markovian character, as attested also by [7], among others. The objective of this paper is therefore to analyze and discuss the performance of the ARRF model for streamflow generation model, that could be applied to hydrological drought assessment in semi-arid regions.

2. Hydrologic Drought Definition

According to [8] four basic considerations must be evaluated for the definition of droughts, that are: 1) what is the greatest interest in the analysis, i.e. what is the nature of the deficit of water to be investigated (meteorological, hydrological or agricultural); 2) what is the time series discretization used in the analysis (annual, semiannual, monthly, etc.); 3) what is the threshold level of separation between flood and drought events; and 4) the choice of the regionalization and standardization methods to be adopted.

To analyze the characteristic parameters of drought and flood events an interval of discretization of one year was adopted in this study, due to the character of the annual events of drought in northeastern Brazil. Furthermore, only the hydrological aspects have been treated herein, i.e. the mathematical modeling of the streamflow series. The mean flow was chosen as a threshold for distinction between flood and drought events, resulting in an average severity for the same flood and drought length periods, within a complete alternating process cycle [9]. Due to the small sample of the annual series, two procedures have been applied for the identification of the probability distribution of the flood and drought duration: decimation and standardization.

A hydrological drought can be defined as a one or more sequenced years, when the average annual flow remains below the long term mean annual flow, considering all the existing series [8]. A drought event may thus be characterized by three parameters, namely the duration D in years; the accumulated deficit or severity S and the magnitude M, which represents the average cumulative deficit below the mean annual flow.

Thus, we may define S, the drought severity as:

$$S = \int_0^D Y(t)dt$$
→ (1)

Where Y(t) = the deficit value of a D-year drought for the year t.
Considering flow rates, we may state:

$$S = \int_0^D [\overline{Q} - Q(t)]dt$$ → (2)

where Q(t) = not regularizable flow rate for year t of a D-year duration drought.
Then we may estimate the magnitude M, as follows:

$$M = \frac{1}{D}\int_0^D Y(t)dt$$ → (3)

or

M = S / D → (4)

Hence, the magnitude may be considered a secondary parameter, since it may be expressed in terms of the duration and the severity, that are considered primary parameters. According to [10], the severity of a D-years duration drought may be described as:

$$S_d = Y_1^D + Y_2^D + ... + Y_D^D$$ → (5)

3. Characteristic Parameters of Drought and Floods Events and Statistical Tests

In order to better understand the phenomenology of the processes involved in modeling, the following statistical analyzes of flood and drought events were performed:
I) Stationarity test (trend);
II) Stochastic test (lag-1 correlation coefficient);
III) Correlation test between the parameters;
IV) Cross correlation between the parameters of successive flood and drought events or inversely.

3.1 Stationary (trend)

Stationarity degrees for flood and drought events were determined by comparing the slope of the regression line adjusted for each series with the t-statistic. The lowest values for the significance level α associated with the rejection of the null hypothesis were estimated (the slope of the regression line equals zero). In this table, NS means that the series in question is very instationary (α <0.01) and S means that the analyzed series is very stationary (α > 0.20). The signals correspond to the signs of the slope of the regression lines.

Table 1: Test t; α values for non-stationary flood and drought events

Station	Flood			Drought		
	Duration	Severity	Magnitud	Duration	Severity	Magnitud
Faz. Cajazeiras	+S	+S	-S	-S	-S	-S
Pacajus	+S	+S	-S	-S	-S	+S
P. Sarasate	-S	+S	+S	-S	-S	-S

1) The signals correspond to the slope of the regression lines.
2) NS (non-stationary) for α <0.01; S (stationary) for α > 020.

Based on the results in Table 1, the following qualitative findings can be made:

I) An increase in the duration of a flood period usually corresponds to an increase in severity and a decrease in the magnitude of flood periods (Figure 1);
II) A reduction in drought duration leads to a decrease in drought severity and an increase in drought magnitude;
III) Neither the characteristic parameters of floods nor drought periods present significant instabilities. This, however, may be due to the fact, that the test used does not have enough power to detect this from a small sample [11].

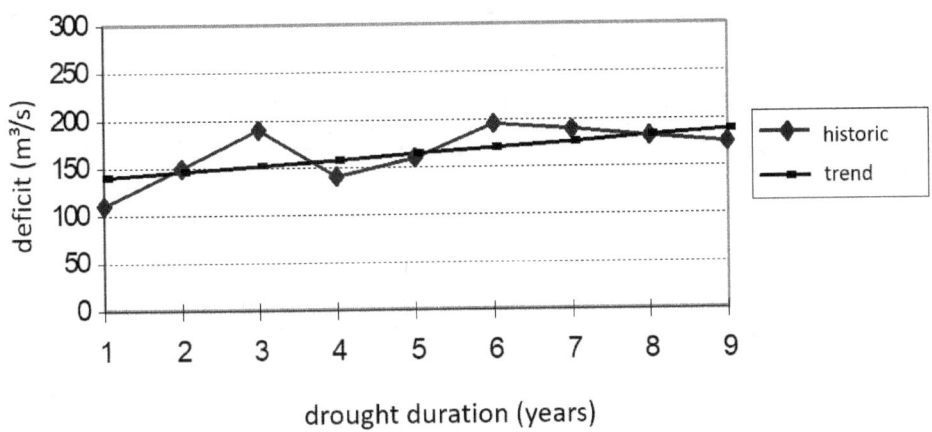

Figure 1: Expected value for the average deficit as a function of drought duration (Faz. Cajazeiras Station)

3.2 Stochasticity

The degree of stochasticity of the series was evaluated by the Anderson test [12] using the lag-1 correlation coefficient, assuming the cyclic series. The lowest significance level α, associated with the null hypothesis, $\rho_1 = 0$, using the z statistic, is shown in Table 2. From this table comes the following qualitative conclusions:

I) Negative correlation coefficients of flood durations are linked to negative correlation coefficients of flood severity and magnitude;
II) Negative correlation coefficients of drought durations are associated with negative correlation coefficients of drought severity;
III) Duration is the most random parameter and the most nonrandom severity;
IV) All characteristic parameters of the flood periods of the tested series presented negative correlation coefficients;

Table 2: Test z; α values for non-random behavior of flood and drought events

Station	Flood			Dought		
	Duration	Severity	Magnitud	Duration	Severity	Magnitud
Faz. Cajazeiras	-R	-R	-R	+0.05	+0.05	+R
Pacajus	-R	-0.10	-0.20	-R	-R	-R
Paulo Sarasate	-R	-0.10	-0.05	+0.05	+0.05	-R

1) The signals correspond to the slope of the regression lines.
2) NR (non-stochastic) for α <0.01; R (stochastic) for α > 020.

The duration and severity of drought events were then through studied envelopes. Estimates of flow maxima were determined by Tschebycheff inequality regardless of probability distribution, as follows:

$$\max X \approx u_x + (n)^{\frac{1}{2}} \sigma_x \quad \rightarrow (6)$$

u_x = mean;
σ_x = standard deviation;
n = sample size.

For random parameters such as duration, [13] proposed to exchange n1 / 2 for (n / 2) 1/2 in the above equation, which is renamed the modified Tschebycheff maximum. Figure 2 shows the dispersion diagram and the envelope for the duration of drought periods for Faz. Cajazeiras Station (Acaraú River) in the state of Ceará.

For non-random parameters, as appears to be the case with severity, severity values are usually restricted to the lower left triangle. This means that periods of extreme drought are only rarely followed by other periods of extreme drought. The envelope thus expresses the maximum response of the watershed to the observed events (figure 3). The upper right triangle therefore has a return period longer than the length of the historical series [13].

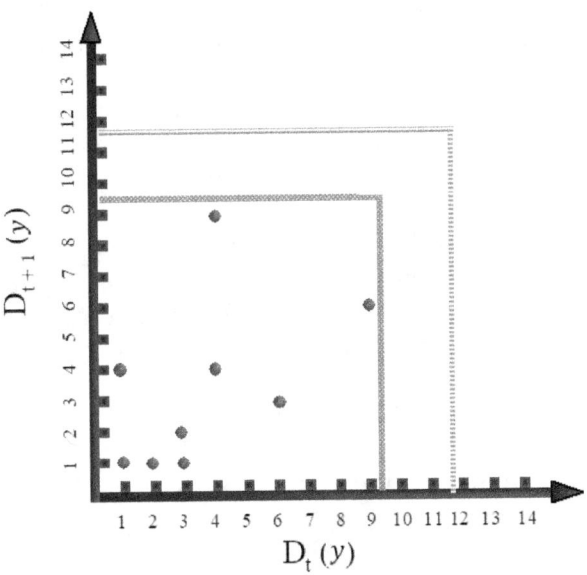

Figure 2: Dispersion diagram, modified Tschebycheff maximum and envelope for the duration of drought periods for Faz. Cajazeiras Station (Acaraú River)

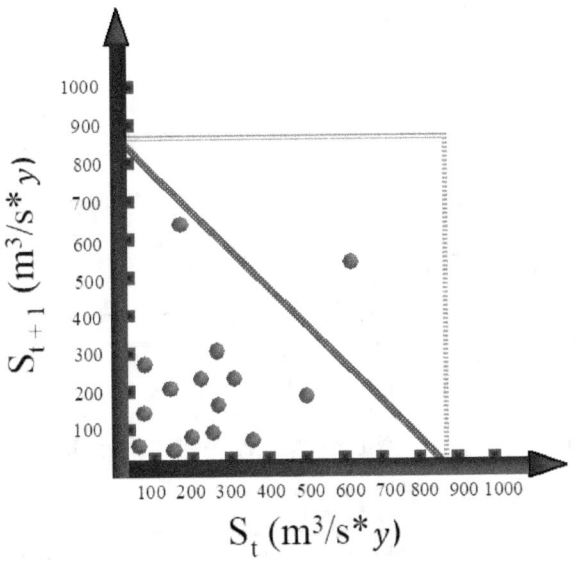

Figure 3: Dispersion diagram, modified Tschebycheff maximum and envelope for the severity of drought periods for Faz. Cajazeiras Station (Acaraú River)

3.3 Simple correlation and cross correlation

The relationship between the three characteristic parameters of hydrological droughts and floods can likewise be studied. We analyzed not only the relationship between each pair of parameters (simple correlation), but also the relationship between flood and drought events or between drought and flood events (cross correlation). This analysis allows a quantification of the internal and external structures of the drought and hydrological flood periods.

Tables 3 and 4 list the significance level values associated with the rejection hypothesis (r = 0). Still, ID means that the correlation is not significant ($\alpha >$ 0.20) and D means that the correlation is more significant ($\alpha <$0.01). The signs correspond to the signs of the coefficients of simple and cross-correlations.

From tables 3 and 4 can be concluded that:

I) Duration and magnitude are the least correlated parameters, while duration and severity are the most correlated.

This internal dependency can be expressed by the following inequality:

$$C_v(M_i) < C_v(D_i) < C_v(S_i)$$

C_v = variation coefficient.

Table 3: Test t; Values of α for simple correlation of flood and drought events.

Station	Flood			Drought		
	D vs. M	D vs. S	M vs. S	D vs. M	D vs. S	M vs. S
Faz. Cajazeiras	+0.10	+D	+D	+0.20	+D	+0.05
Pacajus	+ID	+D	+0.01	-ID	+D	+ID
Paulo Sarasate	+0.05	+D	+D	+ID	+D	+0.10

1) The signals correspond to the slope of the regression lines.
2) ID (uncorrelated) for $\alpha >$ 0.20; D (correlated) for $\alpha <$0.01.

Table 4: Test t; Values of α for the cross-correlation of flood and drought events.

Station	Flood and Drought			Drought and Flood		
	Duration	Severity	Magnitud	Duration	Severity	Magnitud
Faz. Cajazeiras	-ID	-0.10	-0.20	-0.10	-0.20	-ID
Pacajus	+ID	+ID	+ID	+0.20	+0.05	-ID
Paulo Sarasate	-0.20	-0.20	-0.10	-0.10	-ID	+ID

1) The signs correspond to the slope of the cross-regression lines.
2) ID (uncorrelated) for $\alpha >$ 0.20; D (correlated) for $\alpha <$0.01.

From the results of the above tests applied to the flow series one can adopt two procedures for modeling the flow series, namely:
(I) the treatment of duration and magnitude under the assumption of independence;
II) The treatment of duration and severity under the assumption of dependence.

One way of applying this first treatment was presented by [14] by joining a Markov chain for duration with an autoregressive model for magnitude. [8] and [13] demonstrated, however, that the likelihood of extreme droughts is undersized through a Markov process, so that droughts of greater duration and severity than those found in the historical series are rarely generated by this process. An example of a model that follows the second treatment, that is, using the assumption of dependency between duration and severity is the Alternating Renewal-Reward model - ARR, presented by [10]. Here this model was applied to the Brazilian semiarid, associated with the Fragment Method [15], for synthetic monthly streamflow generation, to perform hydrological drought assessment.

4. Alternation Renewal Reward / Fragments (ARRF)

The new model employed is the coupling of the Alternating Renewal Reward model (ARR), shown by [10], used in the generation of annual streamflow values, with the Method of Fragment [15], for the disaggregation in coupled monthly ones.

Annual modeling: Alternating Renewal Reward - ARR

A basic assumption on modeling process of annual flows through the ARR model is that drought events come from different populations, i.e., the deficit Yi (deficit in year i) is uniformly distributed and independent, dependent, however, on duration. For the annual flow generation two stages are performed: 1) drought flood modeling process, 2) flow modeling within drought or flood periods. The model can therefore be found in one of two possible stages. If, for example, the system is assumed a priori to be flood, then DH_1 years of flood are generated. The following step adopted is to assume the system is to be in drought condition, and DL_1 years of drought are generated, and so forth. DH_n and DL_n are intended to have independent and uniformly distributed probabilities.

Thus, the problem consists only in identifying the probability distribution functions for the two-stage model. For the flood/drought modeling process, geometric distributions have been used and for the modeling of flood and drought severity, two-parameter gamma distribution have been applied. A limiting factor for the adjusting of duration (length) and severity distributions is the small sample. During a about 80 years data period, for example, there are approximately only 15 drought periods. To overcome this deficiency two procedures have been employed: decimation and standardization [16].

The decimation procedure [17] was used to obtain n (number of months) streamflow series, using standardized flow values of the straight years, for each drought (flood), to each of the n series. The drought and flood periods severities are thus evaluated. This procedure simulates a regionalization through n flows sets of n different positions of a homogeneous region, subject to the same climatic conditions.

To simulate flood and drought duration geometric distributions was used, as follows:

$$f_x(x) = pq^{x-1} \rightarrow (7)$$

For the severity two-parameter gamma distributions for each duration (time length), was employed, as [10]:

$$f(Y_D) = \frac{\lambda e^{-\lambda Y_D}(\lambda Y_D)^{r-1}}{\Gamma(r)} \rightarrow (8)$$

with

Γ = gamma function; r = shape parameter; λ = scale parameter.

Monthly modeling: Method of Fragment

The Method of Fragments from [15] is based on the disaggregation of annual flows generated by some annually model (in this case, the ARR model) into monthly flows (or shorter time interval). The model is characterized by estimating, for each month j and each year i of the historical flow series, the so-called fragments, given by:

$$f_{i,j} = \frac{Q_{i,j}}{\sum_{j=1}^{n} Q_{i,j}} \rightarrow (9)$$

n = number of months (n = 12); $Q_{i,j}$ = streamflow in month j of year i.

The fragments $f_{i,j}$ correspond to the percentage of annual streamflow (the denominator of the equation above) in year i. Following the historic annual flow values are placed in ascending order and separated into classes. The limits of the class intervals are formed by the mean values of successive flows. The total number of classes is equal to the number of years within the flow series measurement. The first class has zero as the lower limit and the last class upper limit of the last class is infinite. The annual streamflow generated are then distributed according to the class intervals and fragmented into monthly values.

5. Model Application

In order to verify the applicability of the new model to the Brazilian semiarid intermittent rivers, the ARRF model was applied to three basins in the State of Ceará – Brazil, corresponding to the following reservoirs: Paulo Sarasate, Aires de Souza and Carão (Table 5).

Table 5 - Characteristics of the analyzed basins

Station	Reservoir volume (million m^3)	Basin area (km^2)	Average annual flow (m^3/s)	Period
Paulo Sarasate	891	3501	242.0	1912-88
Aires de Souza	104	1092	91.3	1912-88
Carão	20	305	10.9	1912-88

Table 6 presents the estimated values of the parameter p of the geometric distributions adjusted to the three analyzed basins for the drought and flood duration. Figure 4 shows the drought duration curve resulting from adjusting of the geometric distribution, after decimation process for the Paulo Sarasate reservoir basin.

Table 6 - Parameters of the geometric distribution

Station	p-flood (%)	p-drought (%)
Paulo Sarasate	52.5	27.5
Aires de Souza	49.5	35.0
Carão	66.0	22.5

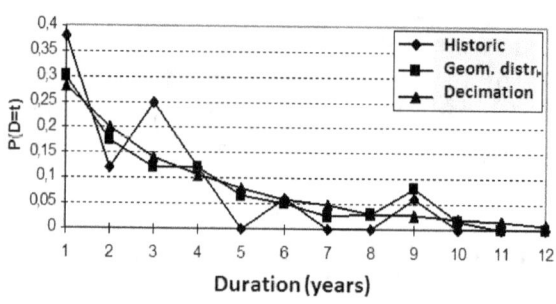

Figure 4 – Drought duration curve for the Paulo Sarasate station

Table 7 contains the gamma distribution estimated parameters for the various flood and drought durations for the Paulo Sarasate station.

For each basin one thousand monthly synthetic streamflow series were generated. Figure 5 shows its mean values. For each month minimum, maximum, median and the historical value have been also shown. Table 8 show the relative average deviations (bias) and the root mean squared errors (RMSE) for the three analyzed basins.

Table 7 - Gamma distribution parameters (Paulo Sarasate station)

Duration (year)	Flood r	Flood λ	Drought r	Drought λ
1	3.212	0.018	38.199	0.306
2	4.653	0.026	35.526	0.299
3	7.052	0.044	27.858	0.236
4	4.101	0.029	47.147	0.373
5	-	-	27.424	0.240
6	-	-	16.418	0.146
7	-	-	14.994	0.138
8	-	-	49.307	0.437
9	-	-	39.137	0.324

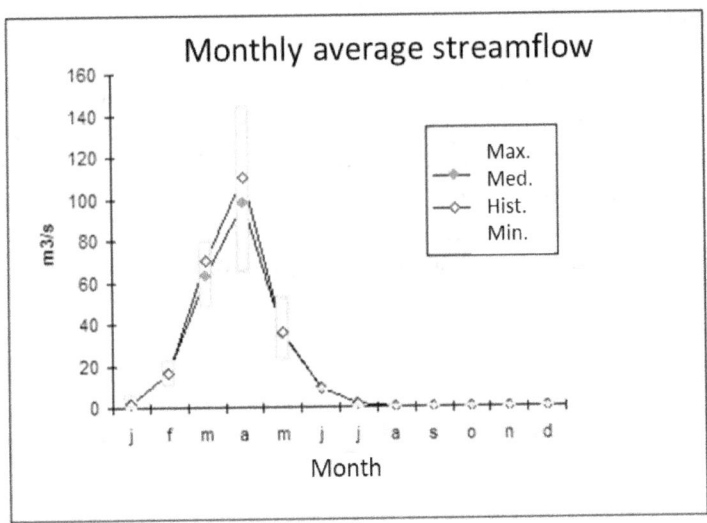

Figure 5 - Generated and historical series: statistical parameters (mean values) for the Paulo Sarasate station

Table 8 - Average adjustment errors (bias and RMSE)

Parameters*	Paulo Sarasate		Aires de Souza		Carão	
	bias	rmse	bias	rmse	bias	rmse
μ	-0.024	0.406	-0.038	0.451	0.005	0.545
σ	-0.025	0.340	0.085	0.465	0.061	0.492
CV	0.053	0.326	0.094	0.395	0.103	0.229
γ	0.007	0.355	0.905	1.358	0.004	0.392
ρ	0.058	0.502	0.119	0.538	-0.009	0.281
ρ lag-1	0.193	0.726	-0.571	0.605	0.340	1.193
* μ = mean; σ = standard deviation; CV = variation coefficient; γ = skewness coefficient; ρ = serial correlation coefficient; ρ lag-1 = lag-1 serial correlation coefficient						

6. Conclusion

This paper presented the application of the ARRF model to generation of synthetic streamflow series in semi-arid regions to using on hydrological drought assessment. This new model, unlike autoregressive models, are based on the characteristics parameter (duration, severity and magnitude) of flood and drought periods, and when applied to different basins of the semi-arid region presents satisfactory results. Thus, it seems to be a useful tool for design and optimization of the operation of reservoir systems in semi-arid regions.

Acknowledgement

This research was conducted partially with the financial support from DAAD (Deutscher Akadmischer Austauschdienst), CAPES (Coordenação de Aperfeiçoamento de Pessoal de Nível Superior) and the Bramar Project – Water Scarcity Mitigation in Northeast Brazil (German-Brazilian research and technology project sponsored by the Federal Ministry of Education and Research – FINEP: Ministério de Ciência, Tecnologia e Inovação).

Conflict of Interest

The authors declare no conflict of interest.

7. References

[1]. FREITAS, M. A. S. - Drought Forecasting, Assessment and Water Allocations Strategies under Climate Change: the 2011-2015 drought period in Northeast Brazil and water crisis of the Cantareira System (São Paulo). European Conference of Tropical Ecology, Göttingen, feb.23-26, 2016.

[2]. MARTINS, E. S.; BRAGA, C.F.C; SOUZA, F.A.S.; MORAES, M.M.G.A.; MARQUES, G.F.; MEDIONDO, E.M.; FREITAS, M.A.S.; VAZQUEZ, V.; ENGLE, N.L.; DE NYS, E. - Climate Change Impacts on Water Resources Manegment: Adaptation challenges and opportunities in Northeast Brazil. Environment and Water Resources Occasional Paper Series, v. 11, n. 1, p. 1-6, 2013.

[3]. ABELS, A.; FREITAS, M. A. S., PINNEKAMP, J., RUSTEBERG, B. - Bramar Project - Water Scarcity Mitigation in Northeast Brazil, 1. ed. Aachen: Department of Environmental Engineering (ISA), 2018, v. 1.,155p., ISBN: 978-3-00-059926-2.

[4]. FREITAS, M. A. S. Tools for Strategic Decision Making on Water Resources Management under Climate Variability and Drought Conditions on the Caatinga's Biome of Northeast Brazil, XXV IUFRO World Congress Forest Research and Cooperation for Sustainable Development - Pesq. flor. bras., Colombo, v. 39, e201902043, Special issue, p. 664-664, 2019.

[5]. FREITAS, M. A. S. - Que Venha a Seca: modelos para gestão de recursos hídricos em regiões semiáridas (Drought Risk Analysis: Models for Water Resources Management in Semiarid Regions), 1ª ed., Rio de Janeiro, 413p, 2010.

[6]. FREITAS, M. A. S.; SILVEIRA, P. B. M.; FREITAS, G. B. - A Resilient Drought Risk Management Approach in the Semiarid Northeast Brazil,

International Journal of Current Research, vol. 11, Issue, 09, pp.6968-6974, september, 2019.

[7]. FREITAS, M. A. S.; FREITAS, G. B. - The GAR(1) Model with Fragment Method for Hydrological Drought Risk Assessment in Semiarid Regions, Brazilian Journal of Development, Curitiba, v. 5, n. 10, p. 18267-18281, sep. 2019 ISSN 2525-8761.

[8]. DRACUP, J. A.; LEE, K. S.; PAULSON, E.G. - On the Definiton of Droughts, Water Resources Research, 16(2), 297-302.

[9]. LEE, K.S.; SADEGHIPOUR, J.; DRACUP, J.A. - An Approach for Frequency Analysis of Multiyear Drought Durations, Water Resources Research, 22(5), 655-662, 1986.

[10]. KENDALL, D. R.; DRACUP, J. A. - On the Generation of Drought Events Using an Alternating Renewal-Reward Model, Stochastic Hydrology and Hydraulics, 6, 55-68, 1992.

[11]. HAAN, C. T. - Statistical Methods in Hydrology, The Iowa State University Press, Ames, Iowa, 1977.

[12]. ANDERSON, R. L. - Distributions of the Serial Correlation Coefficient, Annals of Mathematical Statistics, 13, 1-13, 1942.

[13]. DRACUP, J. A.; LEE, K. S.; PAULSON, E.G. - On the Statistical Characteristics of Drought Events, Water Resources Research, 16(2), 289-296, 1980b.

[14]. JACKSON, B. B. -- Markov Mixture Models for Drought Lenghts, Water Resources Research, 11(1), 64-74, 1975.

[15]. SVANIDZE, G. G. - Mathematical Modeling of Hydrologic Series for Hydroelectric and Water Resources Computations, Water Resources Publications, Fort Collins, Colorado, USA, 1980.

[16]. LEE, K.S.; DRACUP, J.A. - A Stochastic Frequency Analysis of Multiyear Droughts, National Science Foundation 1982, University of California Water Resources Center.

[17]. BLOOMFIELD, P. - Fourier Analysis of Time Series: An Introduction 1976 -- Wiley-Interscience, 258p.

PART II – CLIMATE RESILIENCE IN WATER RESOURCES

"Democracy is based on the hypothesis that everyone can decide about everything. Technocracy, on the contrary, wants to be called upon to decide only those few who have specific knowledge"

Norberto Bobbio
O Futuro da Democracia
Ed. Paz e Terra – 2000

CHAPTER 7

Evaluation of Global Climate Impacts on Water Systems

Marcos Airton de Sousa Freitas[1]

Senior Water Resources Specialist with National Water Agency

Abstract: Climate change in Brazil threatens to intensify difficulties in accessing water. The combination of climate change - in the form of lack of rain or little rain accompanied by high temperatures and high evaporation rates - with competition for water resources, can lead to a potentially catastrophic crisis in several regions. In this chapter, the impacts of global warming on water systems were studied by analyzing the sectors that use water, as well as in the regions of the Amazon, Northeast Brazil, Pantanal and the La Plata Basin.

Keywords: climate change, water systems, global warming impacts.

1. Introduction and Climate Changes

Global warming is causing significant impacts on the environment, such as melting glaciers, changes in temperatures and changes in the amount of precipitation, rising sea levels and warming oceans, among others.

Currently, there is strong evidence about the contribution of human actions to the increasing temperature. The main causes of global warming are the global increases in the concentration of greenhouse gases - mainly carbon dioxide (CO_2), methane gas (CH_4) and nitrous oxide (N_2O) - in the atmosphere due to the use of fossil fuels and change in use land / deforestation.

However, in this chapter, the global impacts of climate change analyzed refer to water systems with respect to the main land components of the hydrological cycle, to user sectors, as well as an analysis of water balance and temperature and rainfall anomalies in the Amazon regions, Brazilian Northeast, Pantanal and Prata Basin.

1.1 Greenhouse Effect and Global Warming

If there were no greenhouse effect, the Earth would be a cold and lifeless planet. The Earth receives solar energy in the form of short-wave radiation.

---

[1] Civil Engineer, Specialist in Water Resources at the National Water Agency - ANA.

Part of this radiation is reflected by the Earth's surface, but most of it is retained in the atmosphere by absorbing this radiation by the greenhouse gases in the atmosphere, such as carbon dioxide, methane, nitrous oxide, among others.

This means that the greenhouse effect is the natural process of the Earth's radiative balance, which allows the retention of sufficient heat in its atmosphere for the maintenance of ecosystems and living beings. However, the Earth is getting warmer. Despite the fact that there is a natural variability of the climate on our planet due to the interactions of biological, physical and chemical factors, the high temperatures and extreme climatic events[2] show a strong link between climate change and human activities.

The IPCC-AR4 (Fourth Assessment Report on Climate Change on the Planet, from the Intergovernmental Panel on Climate Change (Intergovernmental Panel on Climate Change - IPCC) projects an increase in global temperature ranging from 2.0 ºC to 4.5 ºC more than the levels recorded before the Pre-Industrial Era, with a more accurate estimate, an average increase of 3ºC, assuming that carbon dioxide levels stabilize at 45% above the current rate. Previous IPCC, called IPCC-TAR, published in 2001, projected a growth of 1.4ºC to 5.8ºC for 2100.

According to studies and model simulations carried out by IPCC Working Group I, the main causes of global warming are global increases in the concentration of greenhouse gases[3] - mainly carbon dioxide (CO_2), methane (CH_4) and nitrous oxide (N_2O) - in the atmosphere, most likely, largely, from human activities and also by natural factors, such as solar radiation, volcanic activities, oceanic and atmospheric circulation, among others.

The main source of the increase in CO_2 emissions was the use of fossil fuels in the development of economic / daily activities, however, the change in land use, especially due to deforestation, also contributed with a smaller, but significant portion (according to TRA , the proportion of CO_2 emissions is ¾ from fossil fuels and ¼ change in land use). The concentration of CO_2 in the atmosphere in the pre-industrial period was 280 ppm and in 2005 it had increased to 379 ppm.

---

[2] Changes in climate, which include changes in temperatures and Arctic ice, widespread changes in the amount of precipitation, ocean salinity, wind patterns and aspects of extreme weather events such as droughts, heavy precipitation, heat waves and the intensity of tropical cyclones.
[3] Among the best known examples of greenhouse gases are: carbon dioxide (CO_2), methane (CH_4), nitrous oxide (N_2O) and chlorofluorocarbons (CFCs). However, gases such as nitrogen oxides (NOx), carbon monoxide (CO), halocarbons and others of industrial origin such as hydrofluorocarbon (HFC), perfluorocarbon (PFC), water vapor are also examples of gases greenhouse effect.

According to Marengo (2006, p.25 / 26), since the beginning of the Industrial Revolution, in 1750, the atmospheric concentration of carbon dioxide has increased by 31% and, more than half of this growth has occurred in the past 50 (fifty) years. And he complements,

> During the first centuries of the Industrial Revolution, from 1760 to 1960, atmospheric CO_2 concentration levels increased from an estimated 277 parts per million (ppm) to 317 ppm, an increase of 40 ppm. During the recent four decades, from 1960 to 2001, CO_2 concentrations have increased from 317 ppm to 371 ppm, an increase of 54 ppm.

1.1.1 Radiative Forcing

Changes in the amount of greenhouse gases, aerosols or solar radiation are expressed in terms of radiative forcing, a measure of the influence of factors in altering the balance of energy entering and leaving the Earth-atmosphere system.

Positive forcing tends to heat the surface (greenhouse gases), while negative forcing tends to cool it (aerosols, albedo). With regards to negative forcing, there are anthropic contributions to aerosols (sulfate, organic carbon, black carbon, nitrate and dust) and to albedo, which is the property of reflecting radiation any incident.

Deforestation is directly related to the increase or decrease in the albedo of the earth's surface, since the vegetation has a lower albedo than the bare soil, after being deforested. And the lower the albedo, the greater the heat absorption, a fact that would enhance the negative forcing at the expense of the positive.

Another aspect related to radioactive forcing concerns the residence time of gases in the atmosphere. Greenhouse gases (positive forcing) have a long time of existence in the atmosphere, around 50 to 200 years, while aerosols (negative forcing) have a residence time of approximately 1 month.

1.1.2 Effects of global warming

According to IPCC reports (2001-2007), the main effects of global warming are:

a) glaciers melting: Mountain glaciers and snow cover have declined, on average, in both hemispheres. Widespread reductions in glaciers and ice caps contributed to rising sea levels. There are projected increases in melting depth in most subsoil ice regions (permafrost) and a decrease in sea ice in both the Arctic and Antarctica. The defrosting of the ice caps reduces the albedo;

b) changes in temperatures and changes in the amount of precipitation: changes in wind patterns and extreme weather events, such as droughts (increased dry weather and reduced precipitation), heavy rainfall, heat waves and the intensity of tropical cyclones in several regions of the planet. Cold days, cold nights and frosts became less frequent, while hot days, hot nights and heat waves became more frequent. It is very likely that extremes of heat, heat waves and heavy rains will continue to occur more frequently;

c) increase in the level of the oceans and warming of the oceans: salinity of the ocean, with the reduction of the pH culminating in the acidification[4] of the oceans, which result in the dissolution of sodium carbonates and consequent destruction of the corals. Another factor that contributes to the rise of the sea is that the ocean has absorbed more than 80% of the heat added to the climate system, and this causes the sea water to expand, which contributes to the rise in sea level.

The 4th Report of the GT1-IPCC (2007) states that there are still difficulties in the simulation and reliable attribution, on smaller scales, of the observed temperature changes. In fact, there is a universe of variables that involve the presented models, however, the uncertainties tend to decrease, although there is an insufficient database in some regions, mainly in the southern hemisphere.

In addition, the simulation models have greater agreement on the issue of temperature, whereas in the aspects of precipitation and pressure of sea levels there are greater disagreements in the values obtained.

The uncertainty of the results is also linked to the radiative forcing, as it is not yet known what the sensitivity of the forcing is, in order to predict how the weather will react to the components of the radiative forcing.

2. CURRENT CLIMATE

2.1 Global Climate Phenomena

---

[4] The increase in atmospheric concentrations of carbon dioxide leads to an increase in ocean acidification.

The IPCC-AR4 conclusively demonstrates the dangers of increasing the concentration of greenhouse gases in the atmosphere, resulting from the low capacity of industrialized countries to reduce their emissions, as well as the resistance of some developing countries to negotiate stabilization and even decrease in emissions.

According to Marengo et al. (2007), in relation to the causes of climate change, the IPCC-AR4 states that it is "very likely" (up to a 90% chance) that human activities, led by the burning of fossil fuel, have been making the atmosphere warm since mid 20th century. The 2001 report said that this link was "probable" (66% chance or more).

Marengo (2006) states that the global climate models analyzed by the IPCC-TAR show that the warming of the last 100 years is probably not due only to the internal variability of the climate. Assessments based on physical principles indicate that natural forcing cannot explain the observed change in climate in the vertical structure of temperature in the atmosphere.

Marengo et al. (2007) present a summary of the main results of the evaluations made using various levels of climatic and hydrological observations in Brazil and incorporate recent results generated by the PROBIO and GOF-UK projects, which summarize the state of the art of studies developed during the last 2 years research on observed climate variability. The tools commonly adopted to obtain and evaluate past and future climate projections are climate models, called Global Atmospheric Models or Ocean-Atmosphere Coupled Global Models. These models can simulate future climates at a global and regional level in response to changes in the concentration of greenhouse gases and aerosols.

An analysis of the integrated observational evidence for the Brazilian territory points to an increase in average and extreme temperatures in Brazil, both for annual and seasonal values.

According to the data, the Earth is warming up more in the northern hemisphere. In the 21st century, the air temperature globally in 2005 was + 0.48ºC above the average, this being the second hottest year of the observational period, as stated by the Climate Research Unit, from the University of East Anglia, in the United Kingdom.

Regarding precipitation, what has been observed are interdecennial variations of relatively drier or rainier periods in Brazil, in the Amazon and in the Northeast. Regionally, there has been an increase in rainfall in the South and parts of Southern Brazil, in the Paraná - Prata basin, since 1950, consistent with similar trends in other countries in Southeast South America.

In the Southeast, the annual precipitation total does not seem to have undergone any noticeable change in the last 50 years. Therefore, what was observed was that the variations in rainfall and river flows in the Amazon and the Northeast show interannual variability and inter-centennial time scales, which are more important than the trends of increase or decrease.

And that this variability is associated with the phenomena of ENSO (El Niño - Southern Oscillation), the 10-year variability of the Pacific (Pacific Decadal Oscillation - PDO), the Atlantic (North Atlantic Oscillation - NAO), in addition to the variability of the Tropical Atlantic (Atlantic Dipol) and the South Atlantic.

Freitas (1997) presented a simplified scheme for meteorological monitoring, which consists of the application of statistical methods (correlation analysis, contingency tables, analysis of main components, etc.), as well as 'intelligent' systems, such as artificial neural networks, fuzzy logic, etc. for the precipitation forecast. Another possible method would be the forecast, 6 or even 12 months in advance (Chen et al., 1995) of the sea surface temperature (SST) and later through simulation using a global atmospheric-ocean circulation model (Barnett et al., 1994) predicts the total rainfall in a given region. The latter path requires, however, a higher computational cost.

Along with the influence of the positioning of the Intertropical Convergence Zone, among other aspects, several authors reported a possible teleconnection between the phenomenon of El Niño and Southern Oscillation (Southern Oscillation) and rainfall behavior in Northeast Brazil. Stockenius (1981), Ropelewski & Halpert (1987), Rao & Hada (1990), as well as Hastenrath & Greischar (1993) addressed this relationship.

Figure 1 shows the alternation between wet and dry periods from 1911 to 1988 in the State of Ceará (NE-Brazil) and the ENSO years, according to the classification by Rasmusson & Carpenter (1983). The rainfall index used was the LRDI (Lamb Rainfall Departure Index), which expresses in regional terms the deviation of precipitation from the mean, in standard deviation (Lamb et al., 1986).

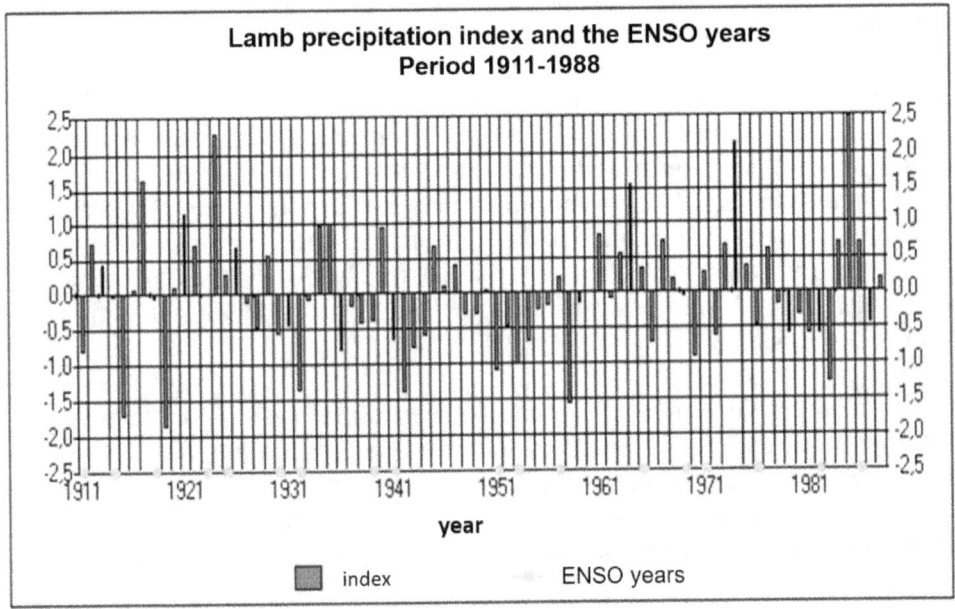

Figure 1: Alternation between wet and dry periods from 1911 to 1988 in the State of Ceará (NE-Brazil) and the ENSO years.

It is easy to see that, in general, years of drought occur after years of the El Niño phenomenon. This happened in the years 1914, 1918, 1930, 1941, 1951, 1953, 1957, 1965, 1969, 1971, 1982 and 1986. The year following an El Niño year, however, is not always a dry year, as can be seen in 1912, 1924, 1926, 1940 and 1977. In addition, there are dry years that have not followed El Niño years, such as 1936 and 1979 (Freitas, 1997).

Positive trends of up to +120 mm / decade are observed in most of the South and Southeast of Brazil, as well as some rainfall stations with negative trends in Amazonas, Bahia, Minas Gerais and Rio de Janeiro. With respect to seasonal precipitation values, the trend of increasing rainfall in southern Brazil is consistent throughout the year, although this trend is more pronounced in the winter months, reaching up to +40 mm / decade and, secondly, during the summer.

For the Northeast, rainfall does not show significant trends of increase or decrease, and in the Amazon, the trends are also not very clear at the regional level. What can be said is that these regions experience inter-dimensional variations, with periods of approximately 25-30 years, alternating rainy seasons. This can be explained by the natural variability of the climate in the form of decadal variations in the Pacific Ocean and the tropical Atlantic (Marengo et al., 2007).

River streamflow analyzes in South America and Brazil (Marengo et al., 2007) point to increases between 2-30% in the Paraná River basin and neighboring regions in Southeast South America, consistent with the analyzes of rain trend in the region. No significant trends were observed in the flows of rivers in the Amazon and in the São Francisco river basin. On the west coast of Peru, the positive trends can be explained by the extremely high values of rainfall and flows during the El Niño years of 1972, 1983, 1986 and 1998 that significantly affect the trends.

In the Amazon, Pantanal and the Northeast, no systematic trends were observed in the long run towards drier or rainier conditions, with interannual and inter-annual variations being more important, associated with natural climate variability in the same temporal scale of variability of inter-phenomena in the Pacific oceans. and tropical Atlantic.

Some of the streamflows in Brazil (Amazonia, Southern Brazil, Northern Northeast) have high correlations with fields of sea surface temperature anomalies in the Pacific and Tropical Atlantic oceans, which suggests a possible association between extreme flows and El Niño or a warming in the North Atlantic Tropical, as was the case, for example, of 1998 with reductions in flows in Manaus and Óbidos and in the low levels of the Solimões River during the recent drought of 2005.

Regarding extreme events, positive trends have been observed in the frequency of hot nights and days and negative trends in the frequency of cold nights and days, consistent with a global warming scenario. For the Southeast of South America an increase in the intensity of episodes and frequency of days with intense rain has been observed in the period 1961-2000, that is, the rains are becoming more and more violent, despite the precipitated annual total. have not undergone any noticeable modification, some studies have shown a relationship between rainfall extremes in southeastern and southern Brazil at frequency / intensity with circulation patterns such as the South Atlantic Convergence Zone (ZCAS) or the South America Low Level Jet (SALLJ).

This is consistent with positive trends of great magnitude in minimum temperatures and to a lesser extent in maximum temperatures in Brazil, as already highlighted in the previous item. Seasonal data show an upward trend in episodes of heavy rain defined by the R10 index (number of days over 10 mm) and the extreme rain extreme index defined by the R95t index (total precipitation fraction due to rain events) above the 95th percentile) in much of the Southeast of South America, Central Argentina and even the Midwest and Southeast of Brazil.

In Brazil, positive trends in rainfall extremes appear more intensely in the states of SP, PR, RS, while the lack of rain data does not allow the analysis to be extended to MG and BA. Positive trends in the maximum rainfall accumulated in 5 days have been observed in the southernmost latitudes of 20 S in South America during the spring, summer and autumn, while some areas of northern Argentina show negative trends during the winter.

With regards to totally atypical extreme events, the highlight of recent years was undoubtedly Hurricane "Catarina", possibly the first hurricane in the South Atlantic, which took residents of the south of the country by surprise in March 2004.

For the South Atlantic there are no reliable statistics on these extreme phenomena, covering a long period of centuries, which could detect other similar phenomena. Therefore, it cannot be said with absolute certainty that Hurricane Catarina was the first event of its kind in the South Atlantic, but certainly nothing comparable has happened in the last 50 years and there are no records in the Brazilian history of such an intense phenomenon on the south coast of Brazil.

According to Mesquita (2005), in the last 50 years a trend has been observed in the Brazilian coast of an increase in the relative sea level, in the order of 40cm / century, or 4mm / year, which may impact around 42 million people , who live in the coastal zone.

The IPCC-TAR (IPCC, 2001) suggests that at the global level, the average sea level may increase between 30 cm and 80 cm, in the next 50 to 80 years. In Brazil, areas most susceptible to erosion are in the Northeast region, due to the lack of rivers capable of supplying the sea with sediments. A 50cm rise in the Atlantic level could consume 100m of beach in the North and Northeast. In Recife, for example, the coastline decreased 80m between 1915 and 1950 and more than 25m between 1985 and 1995. The Rio de Janeiro city is considered one of the Brazilian cities most vulnerable to sea level rise (NAE, 2005).

3. FUTURE CLIMATE

3.1 Climate Scenarios

Marengo et al. (2007) present the projections of future climate scenarios based on the "downscaling" of the HadAM3P model from the Hadley Center in the United Kingdom and the regional climate models developed at CPTEC / INPE and IAG / USP. The projections made for the second half of the 21st century (period 2071-2100) consider the extreme scenarios of high emissions (A2) and low emissions (B2), which were used for the IPCC TAR. Only extreme weather analyzes use the projections generated that are part of the IPCC AR4.

In summary, A1 is the scenario that describes a future world where globalization is dominant (rapid economic growth, small population growth with rapid development of more efficient technologies), with sub-scenarios A1B (stabilization scenario), A1F (maximum use of fossil fuel) and A1T (minimum use of fossil fuel). Scenario A2 describes a heterogeneous future world, where regionalization is dominant. Scenario B1 describes rapid global economic change, where clean technologies are introduced and scenario B2 describes a world in which the emphasis is on local solutions, economic, social and environmental sustainability (Marengo, 2006).

Projections of change in rainfall regimes and distribution, derived from global IPCC models TAR and AR4, for warmer climates in the future are not conclusive, and the uncertainties are still great, as they depend on the models and regions considered. In the Amazon and the Northeast, although some global climate models show drastic reductions in rainfall, other models show an increase. The average of all models, on the other hand, is indicative of a greater likelihood of rain reduction in these regions as a result of global warming. South, Southeast and Midwest do not show any noticeable changes, or a certain increase until the end of the 20th century, but the rains could be more intense. In low latitudes, the projections indicate increases and decreases in the regional continental rain, and one must take into account the high natural variability of the climate (interannual and inter-annual), as is the case of the Amazon and the Northeast (Marengo et al., 2007).

At higher latitudes, the region of the La Plata basin shows projections of possible increases in rainfall and flows up to the second half of the 21st century. This suggests that for this region the future would present a continuity in the variability of rainfall and flows observed during the last 50 years, which perhaps indicates greater confidence in these projections for this region.

There are uncertainties in the observed trends in the variability of climate extremes in Brazil, except perhaps the South region, mainly due to the lack of reliable long-term information or the restricted access to this type of information for large regions, such as the Amazon and Pantanal.

Projections of extreme values for the second half of the 21st century generally show increases in temperature extremes, such as warmer nights, heat waves, and indicators of extreme rain events. Considering the analysis of the global models of the IPCC AR4, they show, for the future, positive trends in the index of hot nights across the continent, which has already been observed in the last 40 or 50 years for some regions of South America and South America. Brazil, and constituting a continuity of the trends observed in the present, although more intense.

In the case of intense rain at present, the trends are positive in the Prata and Northeast Basin, and negative in the south of the Northeast and North of the Amazon. However, for the tropical region, the lack of observations does not allow validating the trends simulated by the models in the present climate, so for the tropical region the uncertainties in the projections of extreme rains are still high.

Climate projections for the second half of the 21st century, for the extreme emission scenarios of IPCC A2 and B2 provide more details on the distribution and intensity of changes in temperature and precipitation in Brazil and South America.

On the continental scale, in relation to precipitation, the region that has the greatest confidence in future climate projections for 2071-2100 is the Northeast, especially during MAM, which is the peak of the rainy season in the North and Northeast. It can be projected that, with medium to high confidence, these scenarios suggest that, in the future, the rainy season in this region may show weaker rains. For the North region, on the other hand, confidence is medium and shows annual and seasonal reductions in rainfall, especially for DJF and MAM. In other regions the sign of change is weaker and with opposite trends between models, and with a low level of confidence. Signal changes tend to be more intense in scenario A2. The other area of positive rain anomalies in the climate of the future, especially during DJF, is the northern coast of Peru - Ecuador, where all regional models show increased rainfall.

Regarding the El Niño - South Oscillation (ENOS) phenomenon, climate projections show little evidence of changes in the magnitude of the phenomenon over the next 100 years. However, there are possibilities for an intensification of the extremes of droughts and floods that occur during hot El Niño events. The projections of the different models indicate that El Niño activity is intense for some, while other models project weaker activity. The uncertainties are derived from the fact that global climate models (which are coupled ocean-atmosphere models) are still unable to correctly simulate the

ENOS phenomenon for the present climate, making their projections of the phenomenon for the climate of the future unreliable.

According to Marengo et al. (2007), studies using simulations of the water balance for the regions of Brazil, considering the projections of temperature and rain of the future climate scenarios generated, suggest in the scenario of higher emissions, a trend of extension of the water deficiency for practically the whole year for the Northeast, which, at present, happens during the dry months, that is, a tendency to "*aridization*" of the semi-arid region until the end of the 21st Century.

For the Amazon, the period of excess water observed in the current climate, during the rainy season, can significantly decrease in future warmer climates, associated with an increase in temperature and evaporation and a reduction in rainfall. The water balances carried out for the present show the Northeast that the rainy and humidity recharge period is between February to April and then the period of water withdrawal and deficiency during the dry season, which runs from July until before the rainy pre-season, in January.

4. CLIMATE IMPACTS ON WATER RESOURCES SYSTEMS

Studies on water systems and the potential for interference from climate change should start from the analysis of historical series. According to Tucci et al. (2003) in Brazil, the hydrological series do not present the necessary data homogeneity due to

> lack of representativeness of historical series to identify the natural variability of climatic processes, changes in the physical / chemical and biological characteristics of the hydrographic basin due to anthropic natural effects and climate changes themselves.

In Brazil, today, it is rare to find historical series older than 80 years and only in the last decades have the number of long series increased throughout the planet, evidencing the interdecennial period of climatic and hydrological processes. Although there are other series of climatic variables that serve as a reference for the analysis of the behavior of the hydrological cycle and the climate (temperatures obtained by correlation with ice or precipitation samples estimated based on tree rings) these provide only an idea of the behavior, but do not allow a more accurate measurement of climatic phenomena.

The risk of using short historical series to analyze and predict the behavior of the water and climate systems is that the samples may not be representative of the wet and dry periods that can be identified in long series. Although there are gaps in the data available for the study of the Brazilian water system in global terms, we have models and scenarios with a good degree of reliability that point to some planetary trends on the influence of climate change on water resources.

According to the Fourth Assessment Report of the Intergovernmental Panel on Climate Change - IPCC of Working Group II (2007) up to the middle of this century, the average annual runoff of rivers and water availability are projected to increase by 10-40% in high latitudes and in some humid tropical areas and decrease by 10-30% in some dry regions in the middle latitudes and in the dry tropics, some of which are already suffering from water scarcity. In some places and seasons, changes differ from these annual values (high confidence).

The aforementioned report (2007) states that it is "likely to increase the extent of drought-affected areas. Heavy rainfall events, the frequency of which is very likely to increase, will increase the risk of flooding. Based on an increasing body of evidence, the IPCC report states that there is high confidence that the following types of hydrological systems are being affected worldwide:

- Increased runoff and anticipation of peak discharge during spring in many rivers fed by glaciers and snow;
- Heating of lakes and rivers in many regions, affecting the thermal structure and water quality."

There is high confidence, based on new and significant evidence, that the changes observed in marine and freshwater biological systems are related to higher water temperatures, as well as the corresponding changes in ice cover, salinity, levels of oxygen and circulation. Among these changes are:

- Displacement of distribution and changes in the amount of algae, plankton and fish in high latitude oceans;
- Increase in the amount of zooplankton in lakes of high latitude and altitude;
- Distribution shifts and anticipated fish migrations in rivers.

The impacts of climate change on hydrology are estimated by defining scenarios of changes in climate "inputs" in a hydrological model of "outputs" of global circulation models (GCMs).

The three key points of this development are: the construction of scenarios that are viable for the analysis of hydrological impacts; the development and realistic use of hydrological models and a better understanding of the interrelations and feedback between the climatic and hydrological systems.

Considerable efforts have been made to develop improved hydrological models to estimate the effects of climate change. Improved models have been developed to simulate water quantity and quality, with a focus on realistic representation of the physical processes involved. These models have often been developed to be of general applicability, with parameters not calibrated locally, and with increasing use of remote sensing data.

Although different hydrological models can generate different flow values for the same input data, the great uncertainties in the effects of the climate on the flows reside in uncertainties in the climate change scenarios, insofar as a conceptual hydrological model is employed. In estimating impacts on aquifer recharge, water quality or floods, however, the transformation of climate in response is less understood, and therefore additional uncertainties are introduced.

In this area, there has been a reduction in uncertainties since the SAR (Second Assessment Report), as the models have been improved and more studies have been carried out. The current impacts on water resources - such as water supply, power generation, navigation, etc. - depend not only on estimated hydrological changes, but also on changes in the demands for resources and in the responses implemented by water resource managers. There have been considerable advances since the SAR in understanding the relationship between the hydrological processes on the earth's surface and the atmospheric processes above that surface.

The ultimate objective of these studies is to improve the analysis of the hydrological effects promoted by climate change using coupled climatic-hydrological models. Some studies have used coupled climatic-hydrological models for flow forecasting (Miller & Kim, 1996) and others to analyze the effects of climate change on flow (Miller & Kim, 2000).

Water resource management is based on minimizing risk and adapting to changing circumstances (usually taking the form of changed demands). A wide range of adaptation techniques has been developed and applied to the sector in recent decades. A very usual classification distinguishes between increased capacity (for example, the construction of reservoirs or structural measures for flood protection), changes in the rules of operation of existing reservoir systems, handling demands and changes in institutional practices.

The first two are known as "supply-side strategies", while the last two are "demand-side strategies". These actions have been implemented regardless of climate change, although changes in water resource management practices have a significant impact on how climate change is expected to impact the sector.

For a more detailed analysis of the impacts of climate change on water resources, it is necessary to study these impacts on the main terrestrial components of the hydrological cycle, such as precipitation, evaporation, soil moisture, aquifer recharge and river streamflow. Of these, precipitation is the main driver of the variability of the water balance in spatial and temporal terms. The hydrological temporal variability in a watershed is influenced by variations in precipitation on a daily, seasonal, annual or decennial scale.

The frequency of floods is affected by changes in variability from year to year in precipitation and changes in short-term rainfall properties, such as the intensity of rain. The frequency of low flows (droughts) and hydrological droughts is primarily affected by changes in the seasonal distribution of precipitation, in the variability from year to year and in the occurrence of prolonged (long-term) droughts.

Regarding rainfall, numerous studies demonstrate different trends in different parts of the world, with a general increase in the middle and high latitudes of the northern hemisphere, particularly in autumn and winter (Carter & Hulme, 1999) and a decrease in precipitation in the tropics and sub-tropics in both hemispheres.

There is evidence that the frequency of extremes has increased in the United States (Karl & Knight, 1998) and the United Kingdom (Osborn et al., 2000). Potential changes in the intensity of precipitation frequency are difficult to infer from global climate models, largely due to large spatial resolution. However, there are indications (Hennessy et al., 1997; McGuffie et al., 1999) that the frequency of intense rain events is generally associated with global warming.

With regards to evaporation, it is known that its rate of the earth's surface is essentially guided by meteorological controls, based on the characteristics of vegetation and soils, and limited by the amount of water available. The properties, type and vegetation cover play an important role in evaporation. A change in the basin's vegetation - directly or indirectly as a result of climate change - can therefore affect its water balance (Friend et al., 1997). As for soil moisture, Komescu et al. (1998) analyzed the implications of climate change for soil moisture variability in southern Turkey, finding substantial reductions in availability during the summer.

Aquifers are fed through effective precipitation, rivers and lakes. A change in the amount of effective precipitation will alter this recharge. The different types of aquifers (confined and unconfined) will be recharged in different ways. Loaiciga et al. (1998) explored the effects of various climate change scenarios on the level of groundwater storage in Texas and demonstrated that in six out of seven scenarios based on global climate models, water levels and exploitation flows would substantially reduce.

By far, most hydrological studies on the effects of climate change fall on potential changes in flow. Georgiyevsky et al. (2001) analyzing eighty European Russian basins, with series ranging from 60 to 110 years in length, found an increase in flows during winter, summer and autumn, from the mid-1970s, and a decrease during spring. Bergstrom & Carlsson (1993), as well as Tarend (1998), found this same pattern for the Baltic region.

In South America, Marengo (1995) found a reduction in flows to most rivers in Colombia from 1970 onwards. In the northwest of the Amazon, Marengo et al. (1998) found an increase in flow for most rivers, starting in the 70s.

For Australia, Thomas & Bates (1997) found a reduction in flow from the mid-1970s to most rivers. Morrison et al. (2002) analyzing the effects of climate change in the Fraser River basin, in Canada, showed that, for the period 2070-2099, the flow predicted by the model resulted in a modest increase of 5% (150 m^3 / s) in the average flow , but a decrease in the peak flow of about 18% (1600 m^3 / s), with these peaks, on average, occurring 24 days earlier in the year. Yu et al. (2002) studying the impacts of climate change on water resources in Taiwan, demonstrated that the flows generated for future climatic conditions were higher for the wet season and lower for the dry season.

In general, it can be said that, for several reasons, it is very difficult to draw quantitative conclusions about the impacts of climate change on water resources. Different studies use different methodologies and different scenarios, but, more importantly, different systems respond very differently to climate change.

It is possible, however, to make some generalizations of a qualitative nature: 1) in systems with large storage capacity, changes in the reliability of water resources may be proportionally smaller than changes in river flows; 2) the potential impacts of climate change need to be considered in the context of other changes that affect water resource management and 3) the implications of climate change are more likely to be greater in systems with the highest water scarcity (IPPC 2001).

4.1 Hydrological Cycle

The projections from the IPCC studies for South America state that the level of uncertainty is still very high regarding rainfall. First, the various numerical simulations do not agree.

According to Nobre (2001)

> Only for the DJF quarter, there is a predominance of simulations indicating increased precipitation. The level of uncertainty regarding changes in the frequency of occurrence of climatic extremes is even greater than for the distribution of rainfall

The rise in temperature in the atmospheric layers closest to the surface will retain more water vapor, which may lead to an "acceleration" of the hydrological cycle, increasing the occurrence of extremes such as strong storms. For Nobre (2001) when the air temperature is higher, there is much more water vapor in the atmosphere. Therefore, in general, a warmer atmosphere with more water vapor will provide more of these extreme phenomena.

In drier areas, climate change and rainfall patterns are expected to lead to salinization and desertification of agricultural land. Future scenarios point to a decrease in some important crops, as well as livestock productivity. It is estimated that changes in precipitation patterns and the hydrological cycle resulting from climate change lead to the disappearance of glaciers, significantly affecting the availability of water for human and animal consumption, agriculture and energy generation.

According to the Stern report "changes in precipitation patterns and extreme weather events will lead to more severe impacts on people than those caused by the warming itself" (Stern Review, 2006) and states that there are more uncertainties about changes in rain regimes in the tropics , mainly because of complex interactions between climate change and natural cycles like El Niño.

4.2 Impacts on User Sectors

4.2.1 Human Water Supply and Sanitation

The greatest vulnerability to climate change to urban supply in Brazil occurs in semi-arid regions, where availability is small, regardless of the regularization provided by dams, in locations supplied by rivers of small basins without regularization, although having a high average flow, and also in urban springs with demand above the capacity of water availability. According to Tucci (2002), another factor that aggravates the scenario of water availability for urban centers is the reduction of availability due to the pollution of water systems.

The greatest weakness of urban supply systems lies in a low level of knowledge on the part of sanitation providers about the water availability of the water sources. The lack of monitoring and long historical series lead to inaccurate information on the availability of the main input from these providers, which can lead to the construction of under-dimensioned reservoirs. The consequence of this is that any major anomaly over the water system can lead to a compromised urban supply.

Considering that the occupation of the Brazilian territory was concentrated on its coastal coast, the number of people impacted by the salinization of surface waters due to the elevated sea level must be significant. Groundwater can also be impacted by a similar process with the infiltration of seawater into the water table.

According to the Atlas of the Semi-arid, prepared by the National Water Agency (ANA, 2006), more than 70% of the cities with a population above 5,000 inhabitants in the northeastern semi-arid (about 1,300 municipalities and 41 million inhabitants) will face crisis in the supply of water for human consumption until 2025.

In general, the impacts of climate change will depend on the affected area and the emissions scenario considered. For example, in the Amazon there may be an increase in temperature of up to 8º C and a reduction in the volume of rain by 20%, if the most pessimistic scenario is confirmed. In the most optimistic scenario, the temperature can rise by around 5º C by 2100.

In global terms, the Fourth Assessment Report of the Intergovernmental Panel on Climate Change (IPCC) of Working Group II, states that some effects of the increase in temperature can already be felt. According to the afore mentioned report (2007, p.11), changes in the planet's climate will affect the health of millions of people, especially those with low adaptive capacity,

Poor communities can be especially vulnerable, particularly those concentrated in high-risk areas. They tend to have more limited adaptation capacities and are more dependent on climate sensitive resources, such as the local supply of water and food.

According to the GTII - IPCC (2007), Latin America is projected that changes in precipitation patterns and the disappearance of glaciers will significantly affect the availability of water for human consumption, agriculture and energy generation. The scenarios analyzed by the IPCC point to an increase in the number of deaths, diseases and injuries due to floods, storms, fires, droughts and heat waves, in addition to the increased consequences of diarrhea that has its means of dissemination in water.

4.2.2 Agriculture

In Brazil, and in general in the world, agriculture is the sector that consumes the most water. About 70% of all the water consumed on the planet goes directly to agriculture, mostly for irrigation. Low water use methods, such as central pivots, are still used on a large scale throughout the country and the incentives to rationalize the use of this resource are still timid in the sector.

According to the contribution of Working Group II to the Fourth Assessment Report of the Intergovernmental Panel on Climate Change - IPCC (2007), some effects of the increase in temperature can already be felt, with medium reliability, in agricultural and forest management in the higher latitudes Northern Hemisphere, such as the anticipation of planting crops and altering the disturbance regimes of forests due to fires and pests.

Crop yields are projected to increase slightly in mid to high latitudes for increases in average local temperature of up to 1 to 3 ° C, depending on the crop, and then decrease in some regions. At lower latitudes, particularly in seasonally dry regions and tropical regions, crop yields are projected to decline even as a result of slight increases in local temperature (1 to 2 degrees Celsius), which would increase the risk of hunger (average confidence - GTII IPCC, 2007).

In drier areas, climate change is expected to lead to salinization and desertification of agricultural land. The productivity of some important crops is projected to decline, as well as livestock productivity, with adverse consequences for food security. In temperate zones, soybean yields are projected to increase (high confidence).

In Brazil, few studies have been done on the reflection of climate change and its impacts on agriculture. A first attempt to identify the impact of climate change on regional production was made by Pinto et al. (1989 and 2001 from CEPAGRI-UNICAMP), where the effects of rising temperatures and rainfall on coffee zoning for the states of São Paulo and Goiás were simulated. There was a drastic reduction in areas with agroclimatic aptitude, condemning production coffee in these regions.

It can be seen that the areas of unfitness for the coffee crop due to the maximum temperatures supported by the plants (23º C annual average) increase significantly until the end of the century, moving the crop progressively to the South and to higher areas, in search for milder weather. The incidence of frost, on the other hand, decreases dramatically.

In southern Brazil, rice irrigation is a major consumer of water, with consumption of 800 people per hectare planted. The tendency is that in a critical scenario of water availability, the conflict between irrigation and water supply will increase. For Tucci et al. (2003), in these situations the intervention will be necessary to enforce the legislation that provides the priority for human supply.

4.2.3 Energy

Brazil is considered one of the countries with the cleanest energy matrix in the world. Today about 80% of the energy produced in the country comes from hydroelectric plants. A weakness of Brazil's current energy production system lies in the fact that the majority of the country's hydroelectric plants have been built in the Southeast, which increases the risk of "blackouts" from the spatial point of view, since a drought in this region it can jeopardize the production of many hydroelectric plants simultaneously.

Since 1970, the Midwest, South and Southeast regions have had an average flow around 30% higher than in the previous period, which means that, for the same installed capacity, it is possible to generate more energy, with less risk of failure. Another proposal to optimize the existing energy generation structure comes from the environmental sector and is based on the repowering of old hydroelectric plants. The proposal seeks to generate energy with low cost and low environmental impact.

According to the document "The Repowering of Hydroelectric Plants as an Alternative to Increase the Energy Supply in Brazil with Environmental Protection" prepared by WWF-Brasil (2004 p.32),

The repowering of old hydroelectric projects has become, now more than ever, the best alternative for power gain for the Brazilian Electric System. The fact that this procedure can be amortized over five years, the environmental impact of execution is low or none, in addition to being a short-term project, justifies this conclusion.

Another way to more efficiently manage energy availability according to Tucci et al. (2002), it is reducing the uncertainties in the forecasts. The forecast of precipitation and its inclusion integrated or separated to the hydrological model will allow to extend the time in advance of the forecast of affluent flows to the reservoirs of the system. Part of this climatic variability has been predicted based on atmospheric circulation models, in horizons of approximately 6 months. These models allow us to predict, with relative success, if the climatic variables will be higher or lower than the climatic average of a season or sequence of months.

Although advances in precipitation forecasting are taking place, it is still impossible to predict long-term climatic conditions, and it is necessary to plan the system, not only so that it can have an emergency plan, but also include premises and practices of energy diversification with a focus on energy. renewable and lease of hydroelectric power plants in order to reduce the risk of system failures.

4.2.4 Navigation

Considering the Brazilian development model based on highways and the current saturation scenario of this road system, other alternatives, which are cheaper and more competitive, are being evaluated. Especially in the North and Center-West regions of the country, where rivers already play a traditional role of displacement channel for the population, some projects for the implementation of waterways have attracted the attention of public policy makers.

Waterway transport, although it has an excellent competitiveness in comparison with other forms of displacement of goods and goods, has a weakness that is dependent on the variability of river levels and its short and medium term forecast, in addition to the statistics of these levels. As most rivers do not have regularization for navigation, the impact of long periods above or below those known as a result of climate change can compromise the price and viability of waterway transport.

4.2.5 Environment

The quality of the water available to the aquatic ecosystem depends directly on the volume of water that passes through the channels of rivers and streams. The greater the flow, the greater the capacity of the water body to dilute the polluting loads and carry out the purification processes necessary for the balance of the ecosystem.

It should be remembered, however, that with the increase in river volumes, the benthic demand (load on the bottom of the rivers) also increases due to a process of erosion of the river channel through the waters. Decomposed and suspended plant matter uses a large amount of oxygen from rivers, reaching levels close to zero, resulting in fish death.

The strategies of damming rivers, aiming to regulate the flow, which can eventually be changed with changes in the climate, can have a serious impact on the fauna and flora of aquatic environments.

According to Tucci et al. (2003, p.17),

> As an environmental constraint for the conservation of fauna and flora, it is much more important to maintain seasonal variability than annually limit values. Obviously, a low limit value can compromise the fauna of the river, but the duration of values above or below certain levels can significantly alter the flora.

Regarding the disposition of ecosystems in Latin America, until the middle of the century, it is projected that increases in temperature and the corresponding reductions in water in the soil will lead to a gradual replacement of the tropical forest by savannah in eastern Amazonia. Semi-arid vegetation will tend to be replaced by arid land vegetation. There is a risk of significant loss of biodiversity due to the extinction of species in many areas of tropical Latin America.

According to the GTII - IPCC (2007, p.09),

> For increases in average global temperature that exceed 1.5 to 2.5 degrees Celsius and corresponding increases in the concentration of carbon dioxide in the atmosphere, major changes in the structure and function of the ecosystem and ecological interactions are projected and geographic distributions of species, with predominantly negative consequences for biodiversity and ecosystem goods and services, such as the supply of water and food (high confidence).

4.3 Impacts on Brazilian Regions

4.3.1 Droughts and Floods

Climate changes resulting from the global increase in temperature due to greenhouse gas emissions are and will cause changes in the rainfall regime in many regions and, consequently, impact water systems. Some processes, such as increasing the evaporation regime, in areas of increased water temperature or deforestation, are already taking place. Others, such as the modification of the rainfall regime with rain prospects, became more intense and with greater interannual variability.

This means that in some regions there will be an intensification of droughts and in others of floods. However, there is still a lot of uncertainty regarding the possible changes in rainfall and changes in the frequency of climatic extremes (droughts, floods, frosts, severe storms, gales, hail, etc.).

Climate change in Brazil threatens to intensify difficulties in accessing water. The combination of climate change - in the form of lack of rain or little rain accompanied by high temperatures and high evaporation rates - with competition for water resources, can lead to a potentially catastrophic crisis in several regions.

Marengo (2006) carried out studies of the impacts of global climate changes for several areas of the Brazilian territory, such as the Brazilian Amazon, Northeast Brazil, Pantanal and the Prata Basin, showing the rain anomalies for the 21st century. The studies indicate great variability in the results when using different global climate models. However, for the 21st century, the data indicate that there is almost always a tendency to increase the temperature although it varies according to the model used.

Salati et al. (2007) presented considerations on the water balance[5] resulting from global climate changes for the Amazon, Northeast Brazil, Paraguay and Prata Basins.

The authors mentioned used, in their analysis, the 5 (five) AOGCM models of the IPCC, being:

1. Hadley Center for Climate Prediction and Research, England (HadCM3);
2. Australia's Commonwealth Scientific and Industrial Research Organization, from Australia (CSIRO-Mk2);
3. Canadian Center for Climate Modeling and Analysis, Canada (CCCMA);

---

[5] It is the difference between water demand and availability in a water body.

4. National Oceanic and Atmospheric Administration NOAA-Geophysical Fluids Dynamic Laboratories (GFDL-CM2);
5. Center for Climate Studies and Research CCSR / National Institute for Environmental Studies NIES (CCSR / NIES).

The analyzes include descriptions of water balances, the seasonal rain cycle for the period 1961-1990 and the future (2050-2100), as well as the anomalies of rain and air and rain temperature in the A2 (High Emission) [6] and B2 scenarios (Low Emission) [7] and in the slicing teams centered on 2020, 2050 and 2080.

4.3.2 Impacts on the semiarid region (Northeast Brazil)

In the Northeast, where water scarcity is already a reality, the impact of climate variability on water resources will be even more dramatic. According to Salati (2007), a large part of the region is semi-arid and has an average rainfall of around 700 mm / year, but losses due to evapotranspiration are above 90%.

The study carried out by the Strategic Affairs Nucleus of the Presidency of the Republic in 2005 (NAE 2005 a, b) suggests that the Northeast is the region most vulnerable to climate change. The Northeastern semi-arid region, which has a short but crucially important rainy season in the present climate, could, in a warmer climate in the future, become an arid region. This can affect regional subsistence agriculture, water availability and population health, forcing populations to migrate, and generating waves of "climate refugees" for large cities in the region or for other regions, increasing social problems already present in large cities.

The results of the analyzes made by Marengo (2006) and Salati (2007) are as follows:

---

[6] It is the scenario that describes a very heterogeneous future world, where regionalization is dominant. There had been a strengthening of regional cultural identities, with an emphasis on family values and local traditions. Other characteristics are high population growth and less concern for rapid economic development.

[7] It is the scenario that describes a world in which the emphasis is on local solutions for economic, social and environmental sustainability. Technological change is more diverse with a strong emphasis on community initiatives and social innovation, rather than global solutions.

Seasonal Rainfall Cycle	Temperature and rainfall anomalies	Water balance
P.124 The models simulate for the 21st century an annual rainfall cycle similar to the present climate. However, the HadCM3 and CCSR / NIES models show less rain during the rainy season and a longer dry season in the A2 and B2 scenarios. The CCCMA and CSIRO models overestimate the rain during the transition from spring to the rainy season.	P.129 For the models used, the future climate tends to be warmer and more humid. Probable consequences: - increased evaporation in the region, caused by high temperatures, which will result in a reduction in the volume of water stored in the soil and a deficit in the water balance. This would not be compensated for by the positive anomalies simulated by the models.	The water balance data made with the average values of the models HadCM3, GFDL, CCCma, SCIRO and NIES, for the two scenarios analyzed (A2 and B2) indicate that there will be a decrease in excess water in the region of up to 100% for the period from 2011 to 2100.

4.3.3 Impacts on the Amazon

The Amazon rainforest has a series of feedback links with climate change that poses a serious threat to the existence of the forest and the continuation of its environmental services.

In this context, for Fearnside (2006, p.397),

> One mechanism is loss of evapotranspiration, thus reducing rainfall to the point where the forest is no longer the type of vegetation favored by the region's climate (for example, Shukla et al., 1990). The forest would be replaced with a type of vegetation similarly to the cerrado, through savanization. Up to 60% of the Amazon rainforest in Brazil could be transformed into cerrado by the process of savanization (Oyama & Nobre, 2003). A separate threat results from the increased frequency of the El Niño phenomenon. El Niño events have increased in frequency since 1976, indicating a change in the global climate system (Nicholls et al., 1996). The El Niño phenomenon causes droughts in the Amazon, which in turn provides conditions for destructive fires, such as those that occurred in Roraima in 1997-1998 (Barbosa & Fearnside, 1999). They also lead to carbon loss from standing forest ecosystems, even in the absence of fire (Tian et al., 1998; Camargo et al., 2004).

For Marengo (2006), extreme climatic events, such as droughts induced by global warming and deforestation, can divide the Amazon in two, and transform Cerrado into an area of 600 thousand km². And the author continues (2006, p.132),

> This "dry Amazon" has vegetation with higher levels of evapotranspiration, and its soils tend to get drier during months without water than soils in very humid regions, and this makes it much more vulnerable to forest fires, the main conversion agent of forest in savanna.

Another fact is that higher temperature averages require that each tree use more water to perform the same amount of photosynthesis.

According to Salati (2007), some published works (Villa Nova et al., 1976 and Salati and Marques, 1984) indicate that deforestation alters the water balance with the decrease in the use of solar radiation for evapotranspiration (latent heat), and with an increase in the use of the radiation balance for heating the soil and the atmosphere (sensitive heat). Thus, deforestation tends to cause a decrease in rainfall and an increase in air temperature.

The results of the analyzes made by Marengo (2006) and Salati (2007) are as follows:

Seasonal Rainfall Cycle	Temperature and rainfall anomalies	Water balance
Os 5 modelos apresentam menos chuva que no clima atual para os cenários A2 e B2.	P.128 Most models have warming in the region + reductions in rainfall, that is, warmer and less humid weather.	For the water balances made with the averages of the values of the models HadCM3, GFDL, CCCma, SCIRO and NIES, for the two scenarios analyzed (A2 and B2) indicate that there will be a reduction of excess water in the region of up to 33% for the period from 2071 to 2100, when comparing the water balance of the period from 1961 to 1990.
Possible consequences p.122 / 123: - duration of the dry season longer than in the current climate; - increased		

susceptibility to large-scale forest fires (combination of high temperatures and drastic reductions in rain in the dry season); - more frequent droughts, especially at the height of the rainy season.		

4.3.4 Impacts on the Pantanal

The Pantanal is one of the most important ecosystems on the planet, representing the largest humid continental area, with approximately 140 thousand km² in Brazilian territory. Its greatest differentiation in relation to other Brazilian biomes is the presence of periodic floods (annual and multiannual), varying from place to place and from year to year, depending on the volume of water and duration. At low tide, the humus-enriched region becomes the largest and richest concentration of natural foods that will sustain all flora and fauna.

The hydrological regime associated with macroclimatic elements and air masses, determines the intensity of the flood. This flood of water from rains and / or overflow from the rivers is due to the low slope of the plain (1 to 2 cm / km in the north-south direction and 10 to 20 cm / km from east to west) and the low depth groundwater, with rapid soil saturation, providing low water runoff.

According to Jorge et al. (2000), studies on the dynamics of atmospheric circulation above the central-west region of Brazil in general, and on the Pantanal in particular are few, certainly making this region one of the least studied from the meteorological point of view.

In this context, they assert that (2000, p.03),

> A particularly important and poorly understood question about the climatology of the central region of Brazil, refers to the origin of the water that precipitates over the region and the destination that it will have after being evaporated. Any element of greater impact, such as the case of fires in central Brazil, can also introduce aerosols into the atmosphere, increasing its local concentration. This influences the physics of clouds and the photochemistry of the atmosphere

above the region in question, factors of considerable relevance in determining long-term climate changes, changes in precipitation patterns and, consequently, in the characteristics of the region's vegetation cover. These important problems remain poorly understood and there are few data available for your studies.

For Marengo (2006), the Pantanal behaves as a gigantic control mechanism for the flooding of the Paraguay River and any significant increase in flow, resulting from climate change or deforestation, will negatively affect the retention and control capacity of this flooded area.

The results of the analyzes made by Marengo (2006) and Salati (2007) are as follows:

Seasonal Rainfall Cycle	Temperature and rainfall anomalies	Water balance
It was not presented.	P.130 The pattern of anomalies is not as coherent as those in the Northeast or the Amazon. The future climate tends to be warmer, but while the models show an intensification of warming up to 2080 for the A2 scenario, some models show increased rainfall and others show reduced rainfall.	It was not presented.

4.3.5 Impacts on the La Plata Basin

It is an important region from the agricultural and hydroelectric generation point of view for the large cities in the Southeast of South America. For Marengo (2006), some impacts caused by climate change may increase the vulnerability of the Prata basin to natural disasters, with floods.

The results of the analyzes made by Marengo (2006) and Salati (2007) are as follows:

Seasonal Rainfall Cycle	Temperature and rainfall anomalies	Water balance
The models simulate for the 21st century an annual rain cycle similar to the	P.131 In general, changes in air temperature are more intense than rain anomalies, and as in the case of	The water balance data made with the average values of the HadCM3, GFDL, CCCma, SCIRO and NIES

present climate. However, the HadCM3 model points to increases in rain in the rainy season and a decrease for the winter season. The CCSR / NEIS and CSIRO models simulate a reduction in rainfall throughout the year.	the Northeast, this increase in air temperature can increase evaporation and compromise the availability of water resources for agriculture and hydroelectricity generation.	models, for the two scenarios analyzed (A2 and B2) indicate that there will be a decrease in excess water in the region for the period from 2011 to 2040 of up to 70% and no excess water for the period from 2041 to 2100, when compared with the water balance data for the period from 1961 to 1990.

5. DISCUSSION AND RESULTS

Climate change in Brazil threatens to intensify difficulties in accessing water. The combination of climate change - in the form of lack of rain or little rain accompanied by high temperatures and high evaporation rates - with competition for water resources, can lead to a potentially catastrophic crisis in several regions.

Climate change in Brazil threatens to intensify difficulties in accessing water. The combination of climate change - in the form of lack of rain or little rain accompanied by high temperatures and high evaporation rates - with competition for water resources, can lead to a potentially catastrophic crisis in several regions.

Regarding the El Niño - South Oscillation (ENSO) phenomenon, climate projections show little evidence of changes in the magnitude of the phenomenon over the next 100 years. However, there are possibilities for an intensification of the extremes of droughts and floods that occur during hot El Niño events.

According to the IPCC report, there is a high confidence that the following types of hydrological systems are being affected worldwide:

- increased runoff and anticipation of peak discharge during spring in many rivers fed by glaciers and snow;
- heating of lakes and rivers in many regions, affecting the thermal structure and water quality."

The rise in temperature in the atmospheric layers closest to the surface will retain more water vapor, which may lead to an "acceleration" of the hydrological cycle, increasing the occurrence of extremes such as strong storms.

Current impacts on water resources - such as water supply, power generation, navigation, etc. - they depend not only on the estimated hydrological changes, but also on changes in the demands for resources and in the responses implemented by water resource managers.

The greatest fragility to climate change for urban supply in Brazil occurs in semi-arid regions, due to low water availability, and in urban springs with demand above the capacity of water availability.

For agriculture, it is projected to increase the productivity of crops in the medium to high latitudes for increases in the average local temperature of up to 1 to 3º C, and in the lower latitudes, especially in the seasonally dry regions and in the tropical regions, it is projected that crop productivity decreases even due to slight increases in local temperature (1 to 2º Celsius), which would increase the risk of hunger (average confidence - GTII IPCC, 2007).

For energy, even though advances in the forecast of precipitation are taking place at the moment, it is impossible to predict long-term climatic conditions, and it is necessary to plan the system not only so that it can have an emergency plan, but also include assumptions and management practices. energy diversification with a focus on renewable energy and leasing of hydroelectric power plants in order to reduce the risk of system failures.

For navigation, as most rivers do not have regularization for navigation, the impact of long periods above or below those known due to climate change can compromise the price and viability of waterway transport.

For the environment, Marengo (2006) and Salati (2007) carried out studies of the impacts of global climate changes for several areas of the Brazilian territory, such as the Brazilian Amazon, Northeast Brazil, Pantanal and the Prata Basin, showing the anomalies of rain and temperature, as well as water balance for the 21st century.

The Northeastern semi-arid region, which has a short but crucially important rainy season in the present climate, could, in a warmer climate in the future, become an arid region.

For the Amazon, most models show warming in the region + reductions in rainfall, that is, warmer and less humid weather. The increase in temperatures causes evapotranspiration, thus reducing precipitation to the point where the forest is replaced with a type of vegetation similarly to the cerrado, through savanization.

In the Pantanal the pattern of anomalies is not as coherent as those in the Northeast or the Amazon. The future climate tends to be warmer, but while the models show an intensification of warming up to 2080 for the A2 scenario, some models show increased rainfall and others show reduced rainfall.

Finally, in the Prata Basin - In general, changes in air temperature are more intense than rain anomalies, and as in the case of the Northeast, this increase in air temperature can increase evaporation and compromise the availability of resources for agriculture and hydroelectricity generation.

6. CONCLUSIONS

Climate changes resulting from the global increase in temperature due to greenhouse gas emissions are and will cause changes in the rainfall regime in many regions and, consequently, impact water systems.

For the GTII - IPCC (2007), Latin America is projected that changes in precipitation patterns and the disappearance of glaciers will significantly affect the availability of water for human consumption, agriculture and energy generation.

The scenarios analyzed by the IPCC point to an increase in the number of deaths, diseases and injuries due to floods, storms, fires, droughts and heat waves, in addition to the increased consequences of diarrhea that has its means of dissemination in water.

There is a greater likelihood of reduced rainfall in the Amazon and Northeast regions as a result of global warming. That said, the trend on the part of the Amazon is "savanization" and the Northeast, "aridization".

The South, Southeast and Midwest regions showed no noticeable changes, pointing out that there are no more in-depth studies on these impacts in the Pantanal, a biome, whose presence of periodic flooding is essential for its existence.

7. BIBLIOGRAPHIC REFERENCES

Barnett, T. P. et al, 1994: Forecasting global ENSO-related climate anomalies, Tellus, 46 a, 361-366.

Bergstrom, S., B. Carlsson, 1993: Hydrology of the Baltic Basin. Swedisch Meteorological ana Hydrological Institute Reports. *Hydrology*, **7**, 21.

Cadernos NAE/ Núcleo de Assuntos Estratégicos da Presidência da República. – nº 3, (fev.2005) – Brasília: Núcleo de Assuntos Estratégicos da Presidência da República, Secretaria de Comunicação de Governo e Gestão Estratégica, 2005, 250 p.

Carter, T. R., M. Hulme, J. F. Crossley, S. Malyshev, m. G. New, M. E. Schlesinger, H. Tuomenvirta, 2000: Climate Change in the 21st Century – Interim Characterizations based on the New IPCC Emissions Scenarios. The Finnish Environment 433, Finnish Environment Institute, Helsinki, 148pp.

Chattopadhyary, N., M. Hulme, 1997: Evaporation and potential evapotranspiration in India under conditions of recent and future climate change. *Agricultural and Forest Meteorology*, **87**, 55-73.

Chen, D., S.E. Zebiak, A. J. Busalacchi, M. A Cane, 1995: An Improved Procedure for El Niño Forecasting: Implications for Predictability, Science, vol. 269, 1699-1702.

Chiew, F. H. S., T. C. Piechota, J. A. Dracup, T. A. McMahon, 1998: El Niño Southern Oscillation and Australian rainfall, streamflow and drought – links and potential for forecasting. *Journal of Hydrology*, **204**, 138-149.

Climate Change 2001 – Impacts, Adaptation, and vulnerability. Contribution of Working Group II to the Third Assessment Report of the Intergovernmental Panel on Climate Change – UNEP (Arnel, Nigel and Liu, Chunzhen) p 195 a 203

Compagnucci, R. H. & W. M. Vargas, 1998: Interannual variability of the Cuyo River's streamflow in the Argentinian Andean mountains and ENSO events. *International Journal of Climatology*, **18**, 1593-1609.

Contribuição do Grupo de Trabalho II ao Quarto Relatório de Avaliação do Painel Intergovernamental sobre Mudanças do Clima – IPCC (2007) – Sumário para formuladores de Políticas. (Adger, Neil e outros)

Fearnside, Philip M. Desmatamento na Amazônia: dinâmica, impactos e controle. In: ACTA AMAZONICA - VOL. 36(3) 2006: 395 – 400 p. Disponível em *http://www.scielo.br/pdf/aa/v36n3/v36n3a18.pdf*

Freitas, M. A. S., 1996: Previsão de Secas por Meio de Métodos Estatísticos e Redes Neurais e Análise de Suas Características Através de Diversos Índices (Ceará -

Nordeste do Brasil), *IX Congresso Brasileiro de Meteorologia*, Campos do Jordão, 6 a 13 de Novembro de 1996.

Freitas, M.A.S., M.H.A. Billib, 1997: Drought Prediction and Characteristic Analysis in Semi-Arid Ceará / Northeast Brazil, Symposium "Sustainability of Water Resources Under Increasing Uncertainty", *IAHS Publ.* No. **240**, 105-112, Rabat, Marrocos.

Friend, A. D., A. K. Stevens, R. G. Knox, M. G. R. Cannell, 1997: A process-based, terrestrial biosphere model of ecosystem dynamics (HYBRID v3.0). *Ecological Modelling*, **95**, 249-287.

Gash, J.; Nobre, C.A.; Roberts, J.M.; Victoria, R., eds.. Amazonian deforestation and climate. New York, John Wiley and Sons, 1996. 611p.

Georgiyevsky, V. Yu, A. V. Yezhov, A. L. Shalygin, 1997: An assessment of changing river runoff due to man's impact and global climate warming. In: River Runoff Calculations, Report at the international Symposium. UNESCO, pp. 75-81.

Hammed et al., 1983: An Analysis of Periodicities in the 1470 to 1974 Beijing Precipitation Record, Geophys. Res. Lett., 10, 436-439.

Hastenrath, S., 1987: The Droughts of Northeast Brazil and their Prediction. In: D.A. Wilhite and W.E. Easterling (ed.) Planning for Drought, Toward a Reduction of Societal Vulnerability, Westview Press/UNEP, Boulder.

Hastenrath, S. & L. Greischar, 1993: Further Work on the Prediction of Northeast Brazil Rainfall Anomalies, Journal of Climate, vol. 6, 753-758.

Hennessy, R. J., J. M. Gregory, J. F. B. Mitchell, 1997: Changes in daily precipitation under enhanced greenhouse conditions. *Climate Dynamics*, **13**, 667-680.

Jorge, Maria Paulete Pereira Martins et al. **Análise Preliminar dos dados de SODAR relativos à camada limite do Pantanal Sul Matogrossense.** 2000, 9 p. Disponível em http://mtc-m15.sid.inpe.br/col/cptec.inpe.br/walmeida/2004/05.27.15.37/doc/Jorge_Analise%20Preliminar.pdf.pdf

Karl, T. R., R. W. Knight, 1998: Secular trends of precipitation amount, frequency and intensity in the United States. *Bulletin of the American Meteorological Society*, **79**, 231-241.

Komescu, A. U., A. Erkan, S. Oz , 1998: Possible impacts of climate change on soil moisture availability in the Southeast Anatolia Development Project Region (GAP): an analysis from an agricultural drought perspective. *Climate Change*, **40**, 519-545.

Lamb, P.J., R.A. Peppler & S. Hastenrath, 1986: Interannual Variability in the Atlantic, Nature, 322, 238-240.

Loaiciga, L. C., D. R. Maidment, J. B. Valdes, 1998: Climate change impacts on the water resources of the Edwards Baclcones Fault Zone aquifer, Texas. ASCE/USEPA Cooperative Agreement CR824540-01-0, American Society of Civil Engineers, Reston VA, USA, 72p.

Marengo, J. A., 1995: Variations and change in South American streamflow, *Climate Change*, **31**, 99-117.

Marengo, J. A., J. Tomasella, C. R. Uvo, 1998: Trenes in streamflow and rainfall in tropical South America: Amazonia, eastern Brazil and northwesterm Peru. *Journal of Geophysical Research – Atmospheres*, 103, 1775-1783.

Marengo, José A. Mudanças climáticas globais e seus efeitos sobre a biodiversidade: caracterização do clima atual e definição das alterações climáticas para o território brasileiro ao longo do século XXI. Brasília: MMA, 2006, 163 p.

McCabe, G. L., 1996: Effects of winter atmospheric circulation on temporal and spatial variability in annual streamflow in the western United Status. *Hydrological Sciences Journal*, **41**, 873-888.

McGuffie, K., A. Henderson-Sellers, N. Holbrook, Z. Kothavala, 1999: Assessing simulations of daily temperature and precipitation variability with global climate models for present and enhanced greenhouse climates. *International Journal of Climatology*, **19**, 1-26.

Mesquita, A. R., A. S. Franco, J. Harari, C. A. Sampaio França, 2005: On sea level along the Brazilian coast – part 2. http://www.mares.io.usp.br/aagn/aagn8/ca/sea_level3_partii.html

Miller, N. L., J. Kim, 1996: Numerical prediction of precipitation and streamflow over Russian River watershed during the January 1995 California storms. *Bulletin of the American Meteorological Society*, **77**, 101-105.

Miller, N. L., J. Kim, 2000: Climate change sensitivity analysis for two California watersheds. *Journal of the American Water Resources Association*, **36**, 657-661.

Morrison, J., M. C. Quick and M. G. G. Foreman, 2002: Climate change in the Fraser River watershed: flow and temperature projections, *Journal of Hydrology*, **263**, 230-244.

Nobre, Carlos - Mudanças climáticas globais:possíveis impactos nos ecossistemas do país, 2001 p.243).

Olsen, J. R., J. R. Stedinger, N. C. Matalas, E. Z. Stakhiv, 1999: Climate variability and flood frequency estimation for the Upper Mississippi and Lower Missouri Rivers. *Journal of the American Water Resources Association*, **35**, 1509-1523.

Osborn, T. J., M. Hulme, P. D. Jones, T. A. Basnet, 2000: Observed trends in the daily intensity of United Kingdom precipitation. *International Journal of Climatology*, **20**, 347-364.

Piechota, T. C., J. A. Dracup, R. G. Fovell, 1997: Western U. S. streamflow and atmospheric circulation patterns during El Niño-Southern Oscillation. *Journal of Hydrology*, **201**, 249-271.

Rao, V. B. & K. Hada, 1990: Characteristics of Rainfall over Brazil: Annual Variations and Connections with the Southern Oscillation, Theor. Appl. Cilmatol., 42, 81-91.

Ropelewski, C. F. & M. S. Halpert, 1987: Global and Regional Scale Precipitation Patterns Associated with the El Niño / Southern Oscillation, Mon. Wea. Ver., 115, 1606-1626.

Salati, Eneas et al. Tendências das Variações Climáticas para o Brasil no Século XX e Balanços Hídricos para Cenários Climáticos para o Século XXI. Brasília: MMA, 2007, 186 p. Disponível em http://www6.cptec.inpe.br/mudancas_climaticas/prod_probio/Relatorio_4.pdf

Schulze, R. E., 1997: Impacts of global climate change in a hydrologically vulnerable region: challenges to South African hydrologists. *Progress in Physical Geography*, 21, 113-136.

Shorthouse, C., N. W. Arnell, 1997: Spatial and temporal variability in European river flows and the North Atlantic Oscillation. FRIEND'97: *International Association of Hydrological Science Publications*, **246**, 77-85.

Stockenius, T., 1981: Interannual Variations of Tropical Precipitation Patterns, Mon. Wea. Ver., 109, 1233-1247.

Tarend, D. D., 1998: Changing flow regimes in the Baltic States. In: Proceedings of the Second International Conference on Climate and Water, Espoo, Finland, August, 1998. Helsinki University of Tecnology, Helsinki, Filand.

Thomas, I. F., B. C. Bates, 1997: Responses to the variability and increasing uncertainty of climate in Australia. *The Third IHP/IAHS G. Covach Colloquium: Risk, Reliability, Uncertainty and Robustness of Water Resources Systems*, 19-21 September, 1996. UNESCO, Paris, France.

Tucci, Carlos et al. Clima e Recursos Hídricos no Brasil (2003 – coleção ABRH 9) 1ª edição -Porto Alegre –RS p.1 a 17

Vogel, R. M., C. J. Bell, N. M. Fennessey, 1997: Climate, streamflow and water supply in the northeastern United States. *Journal of Hydrology*, **198**, 42-68.

Wilby, R. and T. M. L. Wigley, 1997: Downscaling general circulation model output: a review of methods and limitations, *Progress in Physical Geography*, **21**, 530-548.

Wildby, R., T. M. L. Wigley, D. Conway, P. D. Jones, B. C. Hewitson, J. Main, and D. S. Wilks, 1998: Statistical downscaling of general circulation model output: a comparison of methods. *Water Resources Research*, **34**, 2995-3008.

Wildby, R., L. E. Hay, G. H. Leavesley, 1999: A comparison of downscaled and raw GCM output: implications for climate change scenarios in the San Juan River Basin, Colorado. *Journal of Hydrology*, **225**, 67-91.

Yu, Pao-Shan, Tao-Chang Yang and Chih-Kang Wu, 2002: Impact of climate change on water resources in southern Taiwan, *Journal of Hydrology*, **260**, 161-175.

CHAPTER 8

RESERVOIR OPERATION RULES AS MITIGATION AND ADAPTATION ACTIONS: THE CASE OF PARAIBA DO SUL SYSTEM

Marcos Airton de Sousa Freitas

Senior Water Resources Specialist with National Water Agency

ABSTRACT: The Paraíba do Sul River basin, where 5 million inhabitants live, has about 57,000 km^2 and extends across the states of São Paulo, Rio de Janeiro and Minas Gerais. Approximately 2/3 of its waters are transferred to the Guandu basin, supplying another 8 million inhabitants of the metropolitan region of Rio de Janeiro, in addition to producing energy and supplying water to the region's industries. On July 17, 2000, through Law No. 9,984, the National Water Agency - ANA was created, having among its attributions, to define the operating conditions for hydroelectric reservoirs in conjunction with the National Electric System Operator - ONS. The aim of this article is to make a synthesis of the simulations carried out, and later discussed within the scope of the Basin Committee, resulting in the edition of Resolution No. 211/2003, providing for the rules to be adopted for the operation of the Paraíba River hydraulic system do Sul, which comprises, in addition to the reservoirs located in the basin, also the structures for transposing the waters of the Paraíba do Sul River to the Guandu System. The rules in force until then had been established by the DNAEE Ordinance No. 022/1977 and by Decree No. 81.436 / 1978. It was intended to present the Decision Support System developed for the simulation of said system, using the flow network model, called AcquaNet, used in watershed simulation.

KEYWORDS: reservoir operating rules; decision support system; Paraíba do Sul.

1. INTRODUCTION

On July 17, 2000, through Law No. 9,984, the National Water Agency - ANA was created, having among its attributions, to define the operating conditions for hydroelectric reservoirs in conjunction with the National Electric System Operator - ONS.

Since its implementation in December 2000, ANA has been conducting studies with a view to reassessing the rules currently used by ONS for the operation of the Paraíba do Sul basin reservoirs (Figure 1), which is based on the following legal provisions:) Decree No. 68,324, of March 9, 1971; b) Decree No. 73,619, of February 12, 1974, designating DNAEE as responsible for establishing the operating rules of the reservoirs; c) DNAEE Ordinance No. 022, of February 14, 1977, defining GCOI as responsible for monitoring the Paraíba do Sul river operation. From this monitoring, GCOI should submit the new operating rules to DNAEE for approval; d) Decree No. 81,436, of March 9, 1978.

Since 1997, the Paraibuna (drainage area of 4,150 km²) and Santa Branca reservoirs (drainage area of 5,030 km2), located on the Paraíba do Sul river, and the Jaguari reservoir (drainage area of 1,300 km²), on the river Jaguari, had not been able to recover their useful volumes. The Paraibuna and Jaguari reservoirs, considered the "lungs" of the hydraulic system of the Paraíba do Sul river hydrographic basin, as they together represent about 80% of the total storage of the basin.

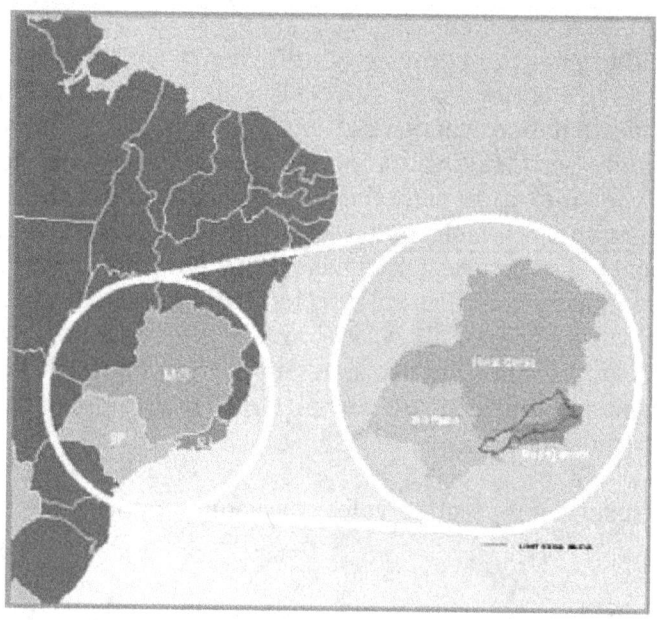

Figure 1: Location of the Paraíba do Sul River Basin

Figure 2 shows the schematic of the hydraulic system of the Paraíba do Sul River and its transposition to the Guandu River basin.

Figure 3 shows the evolution of the equivalent storage, of the basin reservoirs as a percentage of the useful volume, starting in January 1993. The limit curve for reducing the objective flow in Santa Cecília from 250 m³/s to 190 m³/s, established by Ordinance DNAEE 022, of 1977.

The annual evolution of equivalent storage and the forecast for the evolution of this storage for the end of the dry period in 2003 are shown in figure 4.

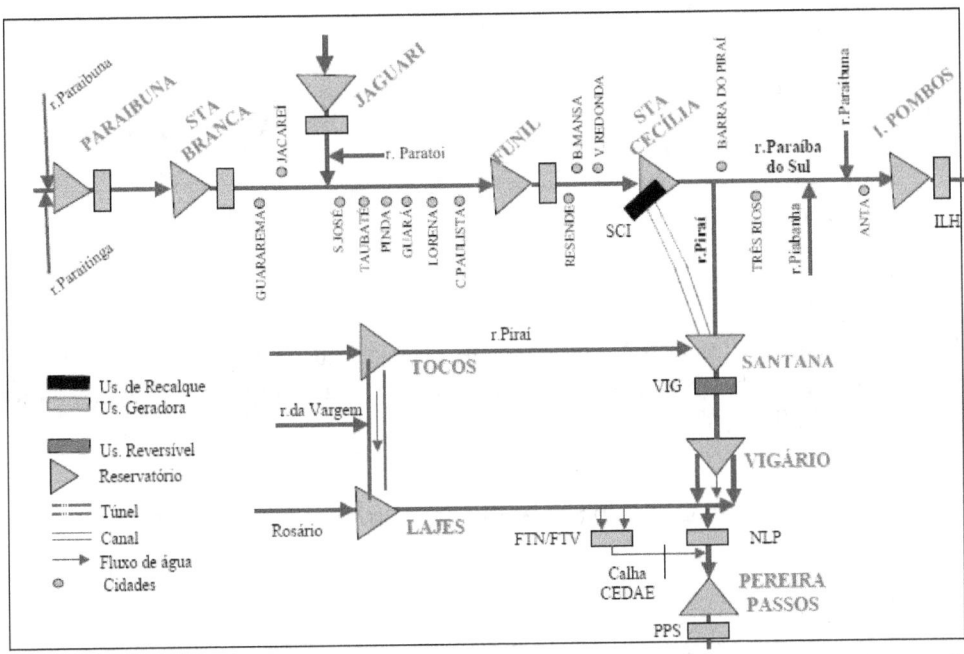

Figure 2: Hydraulic System Scheme of the Paraíba do Sul River and Guandu River Basin Transposition.

The reason for the reduced affluent flows could be associated with the low rainfall in recent years in the hydrographic basin, with the inflows being above the capacity to regulate the reservoirs, or a combination of both factors. In this sense, an assessment was made of the regularization capacity of the Paraibuna and Jaguari reservoirs, considered the "lungs" of the hydraulic system of the Paraíba do Sul River hydrographic basin. The flow series available in the Electric Potential Information System were adopted. Brazilian - SIPOT. The historical evolution of the inflow flows to the Paraibuna and Jaguari reservoirs for the period from 1993 to 2003 were also analyzed.

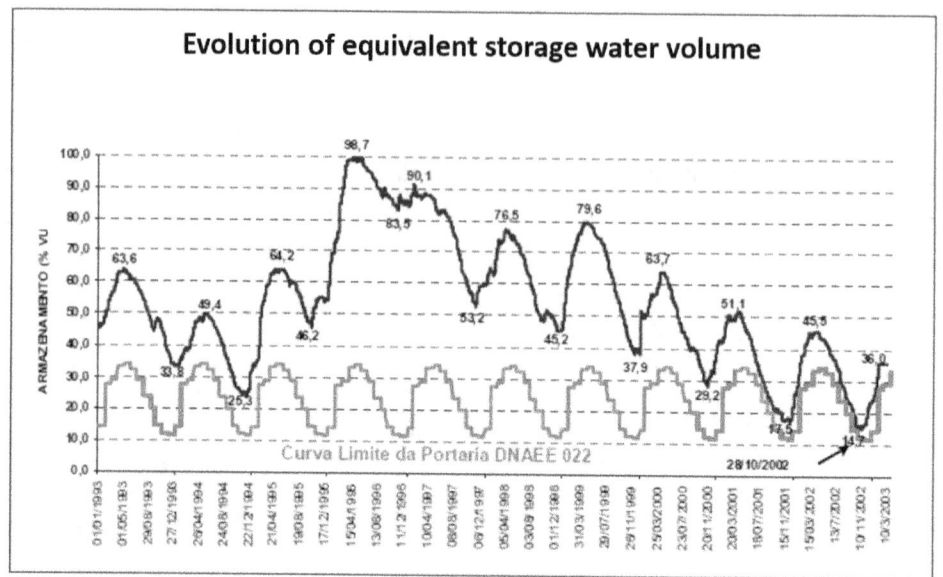

Figure 3. Evolution of equivalent storage water volume in the Paraíba do Sul basin together with the limit curve for reducing the objective flow in Santa Cecília from 260 m³/s to 190 m³/s.

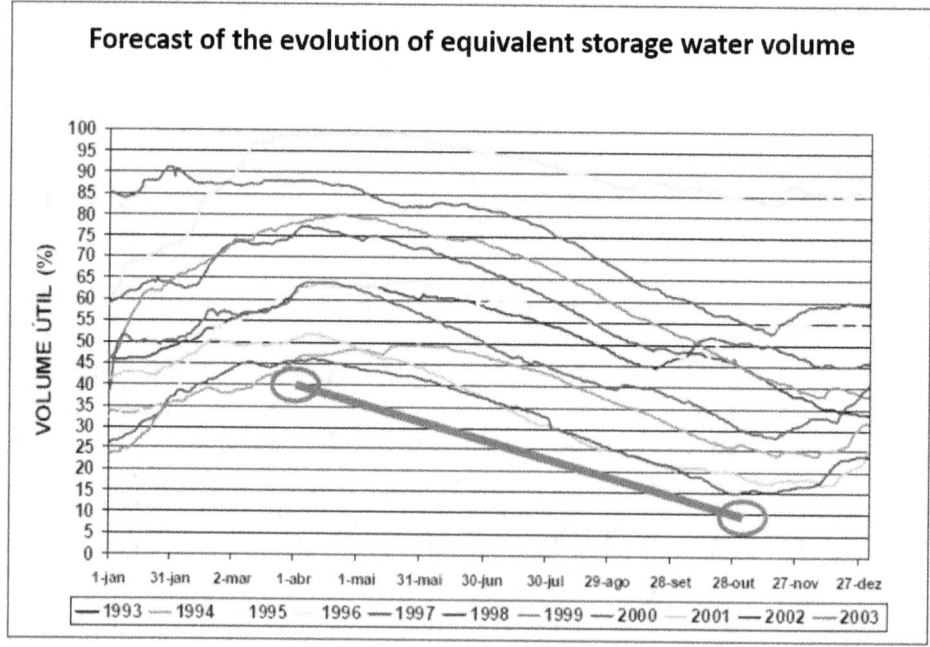

Figure 4. Forecast of the evolution of equivalent storage water volume in the Paraíba do Sul basin.

2. EVOLUTION OF STORED VOLUMES IN THE BASIN'S MAIN RESERVOIRS

Table 1 shows the main characteristics of the Paraibuna and Jaguari reservoirs.

Table 1. Characteristics of the Paraibuna and Jaguari reservoirs.

hydroelectric power plant	date of beginning of filling of the reservoir	date of commencement of operation	maximum operational water level	minimum operational water level	total volume (10 hm³)	useful volume (10 hm³)	long-term average flow (m³/s)
Paraibuna	january 1974	april 1978	714.00	694.60	4,732	2,636	69.00
Jaguari	december 1969	may 1972	623.00	603.20	1,236	793	28.00

Figures 5 and 6 show the evolution of the average monthly storage of the reservoirs, since 1993.

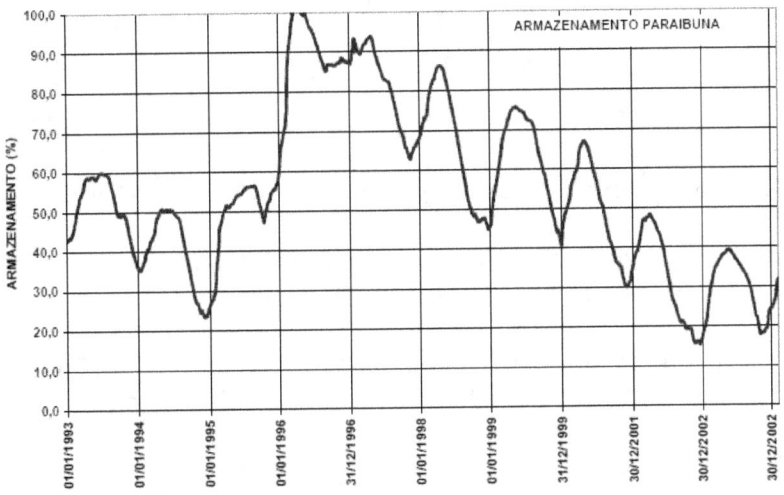

Figure 5. Evolution of storage volumes in the Paraibuna reservoir

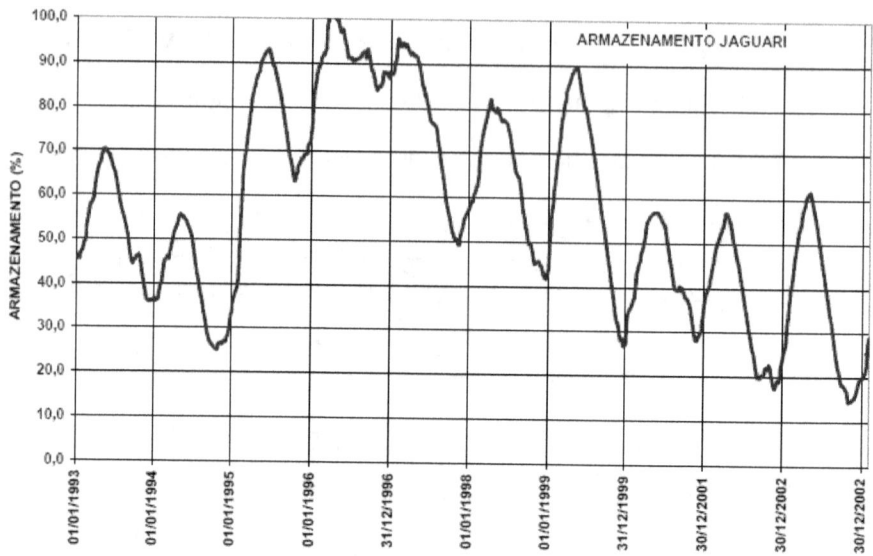

Figure 6. Evolution of storage volumes in the Jaguari reservoir

As can be seen in these figures, from 1997 onwards the reservoirs were unable to recover their useful (operational) volumes.

Table 2 shows the main minimum flow requirements (Qmin) and maximum flow rates (Qmax) established in the legal provisions in force until 2002.

Table 2. Minimum and maximum flow rate requirements

	Law n. 68.324/1971	DNAEE Ordinance n. 022/1977	Law n. 81.436/1978
Downstream Paraibuna		Qmin = 30 m³/s	
Downstream Santa Branca		Qmin = 40 m³/s	
Downstream Jaguari		Qmin = 10 m³/s; Qmin = 42 m³/s in the dry period (June to November)	
Downstream Funil		Qmin = 80 m³/s	
Upstream Santa Cecília		Qmin = 190 m³/s	
Downstream Santa Cecília	Qmin = 90 m³/s	Qmin = 90 m³/s	Qmin = 90 m³/s; Qmin = 71 m³/s in adverse hydrological conditions
Pumping in Santa Cecília	Qmax = 160 m³/s	Qmin = 100 m³/s	Qmin = 119 m³/s in adverse hydrological conditions (Q=190-71)

3. RAINFALL, AFFLUENT FLOWS TO THE MAIN BASIN RESERVOIRS AND ESTIMATED DEMAND

The Paraibuna (Paraibuna / Paraitinga River) and Jaguari (Jaguari River) plants, located at the head of the basin, without, therefore, impacting the operation of upstream reservoirs, were chosen for a first analysis. The series of natural monthly flows, used in the Brazilian Electric Potential Information System (SIPOT, 2003), were used.

Figures 7 and 8 show the historical evolution of natural flows, from 1931 to 2002, respectively, for the Paraibuna and Jaguari plants. As can be seen, since 1997 natural flow rates have been below the average long-term flow. Table 3 shows the flow rates captured between the reservoir stretches, estimated by the National Water Resources Plan.

Table 3. Flow rates captured between sections of reservoirs by the National Water Resources Plan.

Vazão captada (m³/s) - Plano Nacional

Demanda	Paraibuna	Santa Branca	Funil	Santa Cecília	Jaguari	Total
Humana	0.30	0.09	6.21	2.67	2.86	12.13
Animal	0.07	0.02	0.19	0.09	0.02	0.39
Indústria	0.11	0.07	4.59	1.22	2.46	8.45
Irrigação	0.14	0.02	7.45	0.26	0.13	8.01
Total	0.62	0.20	18.44	4.24	5.47	28.97

The rainfall series analyzed did not show, statistically, a reduction in precipitation in the period from 1997 to 2002.

However, the average precipitation in the Paraíba do Sul river basin shows a decrease in rainfall totals from October 2001. According to INPE, the total accumulated in the period from October 2001 to April 14, 2003 was 2019mm and the long-term average for the same period was 2676mm. Figure 7 shows this decrease monthly.

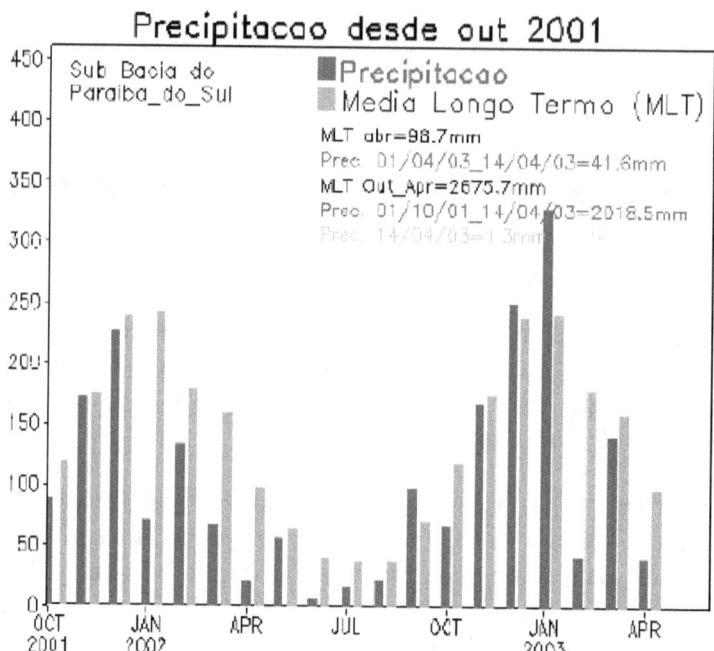

Figure 7. Average Monthly Precipitation in the period October / 2001 to April / 2003 and comparative with the Long Term Average for the Paraíba do Sul River basin. Source: (http://www.cptec.inpe.br/%7energia)

4. DEFLUENT FLOWS TO THE MAIN RESERVOIRS

In DNAEE Ordinance No. 022, of February 14, 1977, Qmin = 30 m^3/s was established downstream from Paraibuna. For the Jaguari reservoir, Qmin = 10 m^3/s and Qmax = 42 m^3/s in the annual dry period (June to November).

The historical evolution of the flows flowing into the Paraibuna and Jaguari reservoirs, respectively, for the period from 1993 to 2003 are shown in figures 8 and 9. For the Paraibuna reservoir, it was observed, especially in the period of 2001, that there was a defluent flow greater than the minimum value of 30 m^3/s practically throughout the year, except for the months of January and February.

In 2002, the minimum defluent flow of 30 m^3/s was practiced during the first five months. For the other months, the defluent flow was greater. Regarding the Jaguari reservoir, for the years 2001 and 2002, DNAEE Ordinance No. 022, of February 14, 1977, was fulfilled, which establishes a minimum flow of 10 m^3/s and a maximum flow of 42 m^3/s, for the months of June to November.

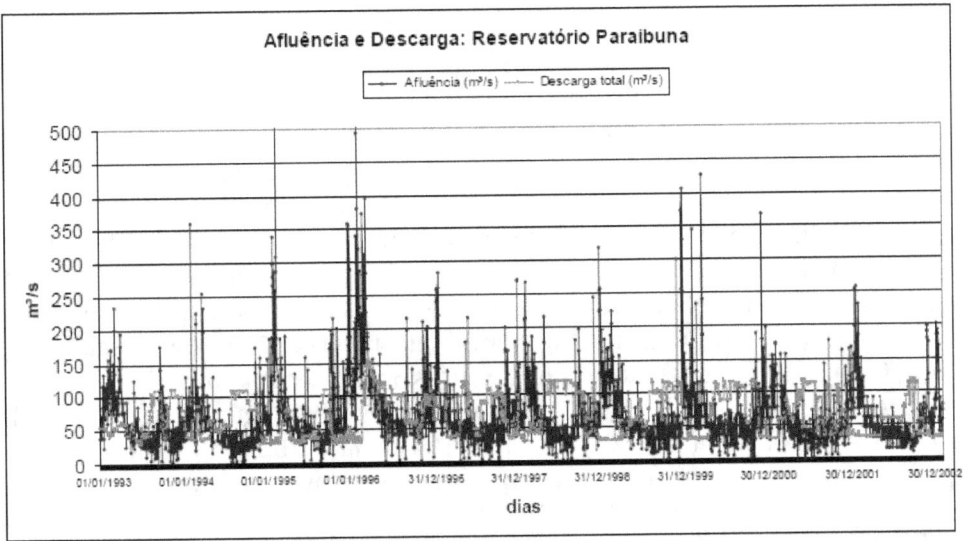

Figure 8. Affluence and discharge. Paraibuna Reservoir (1993 to 2003).

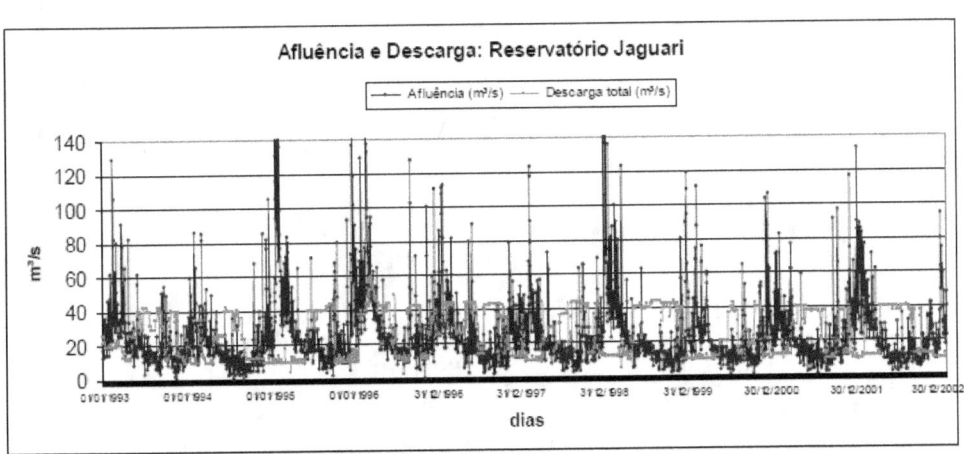

Figure 9. Affluence and discharge. Jaguari Reservoir (1993 to 2003).

5. DECISION SUPPORT SYSTEM FOR THE PARAIBA DO SUL BASIN

To simulate the operation of the Paraíba do Sul basin reservoirs, the hydraulic system of the basin was used, using the AcquaNet model (figure 10). AcquaNet is a flow network model for watershed simulation. With it, you can set up networks with large number of reservoirs (represented by triangles), demands (squares) and stretches of channels (connections), representing the problem under study in a more detailed way. The circles or nodes of passages represent the confluences of rivers, pumping stations or hydroelectric plants. AcquaNet was developed by LabSid from Escola Politécnica da USP and it originates from the MODSIM model, developed at the University of Colorado (Labadie, 1988).

Figure 11 shows the results of the reservoir simulations. The "steady" or regularized flows by the Paraibuna and Jaguari reservoirs are shown, respectively, for different levels of guarantee, considering the historical inflow series from 1931 to 2002. For the Paraibuna reservoir, the flow regularized with 100% guarantee (Q100% = 62.78 m^3/s) represents about 90.97% of the average long-term flow (69.01 m^3 / s). For the Jaguari reservoir, the regularized flow with 100% guarantee (Q100% = 25.2 m^3/s) corresponds approximately to 87.74% of the average long-term flow (28.72 m^3/s).

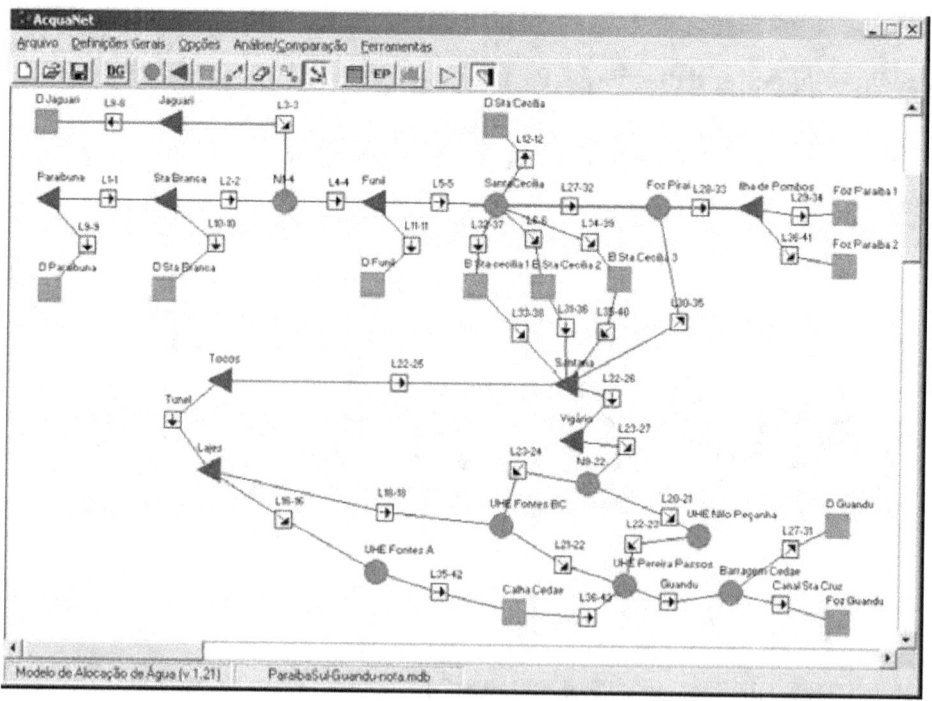

Figure 10. Representation of the operation model of the Paraíba do Sul River basin reservoirs.

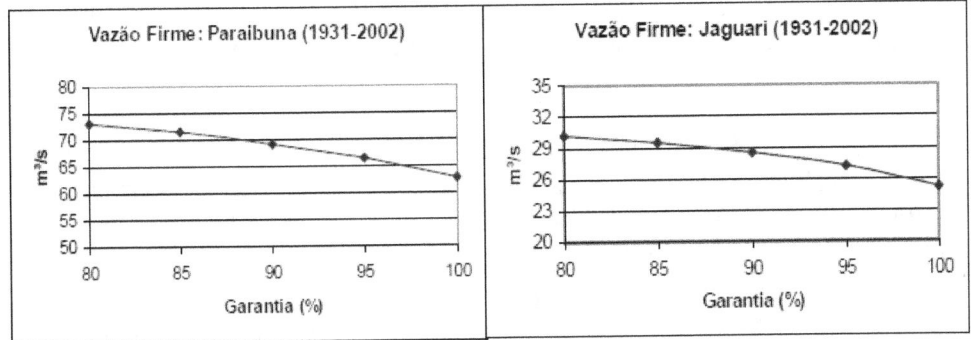

Figure 11. Firm flow of the Paraibuna reservoir and Jaguari reservoir

5.1 New Operation Rules for Basin Reservoirs

With the use of the model implemented in AcquaNet, it is possible to test innumerable possibilities of inflow flows to each reservoir, as well as restrictions of maximum values and order of priority in the depletion of the reservoirs. After articulation with the National Electricity System Operator - ONS and with the Paraíba do Sul River Basin Committee - CEIVAP and Guandu River Basin Committee, Resolution No. 211, of May 26, 2003, was published, which provides for the rules to be adopted for the operation of the Paraíba do Sul River hydraulic system, which includes, in addition to the reservoirs located in the basin, also the structures for transposing the waters of the Paraíba do Sul River to the Guandu system.

Article 1 of the afore mentioned Resolution establishes, on an emergency basis, the following operating rules for the hydraulic system of the Paraíba do Sul River:

I - the minimum discharge downstream of the facilities must respect the following limits:
a) Paraibuna 30 m^3/s
b) Santa Branca 40 m^3/s
c) Jaguari 10 m^3/s
d) Funll 80 m^3/s
e) Santa Cecília 71 m^3/s (instantaneous)
f) Pereira Passos 120 m^3/s (instantaneous);

II - when the incremental flow between Funil and Santa Cecília is greater than 110 m^3/s, the emergency flow of 71m^3/s downstream from Santa Cecília should be gradually increased until reaching the limit of the minimum normal flow of 90m^3/s;

III - the minimum limit for the average pumping flow in Santa Cecília is $119 m^3/s$;

VI - the depletion of reservoirs to meet the minimum limit of 190 m^3/s in Santa Cecília (71 m^3/s for downstream and 119 m^3/s for pumping) must observe the following order of priority, trying to maintain the limit of 10% their useful volume: a) 1st - Funil; b) 2nd - Santa Branca; c) 3rd - Paraibuna; d) 4th - Jaguari.

The sole paragraph of Art. 1 indicates that the order of priority of depletion may be revised, depending on the inflows usually verified to avoid a sharp imbalance between the storage of the Paraibuna and Jaguari reservoirs.

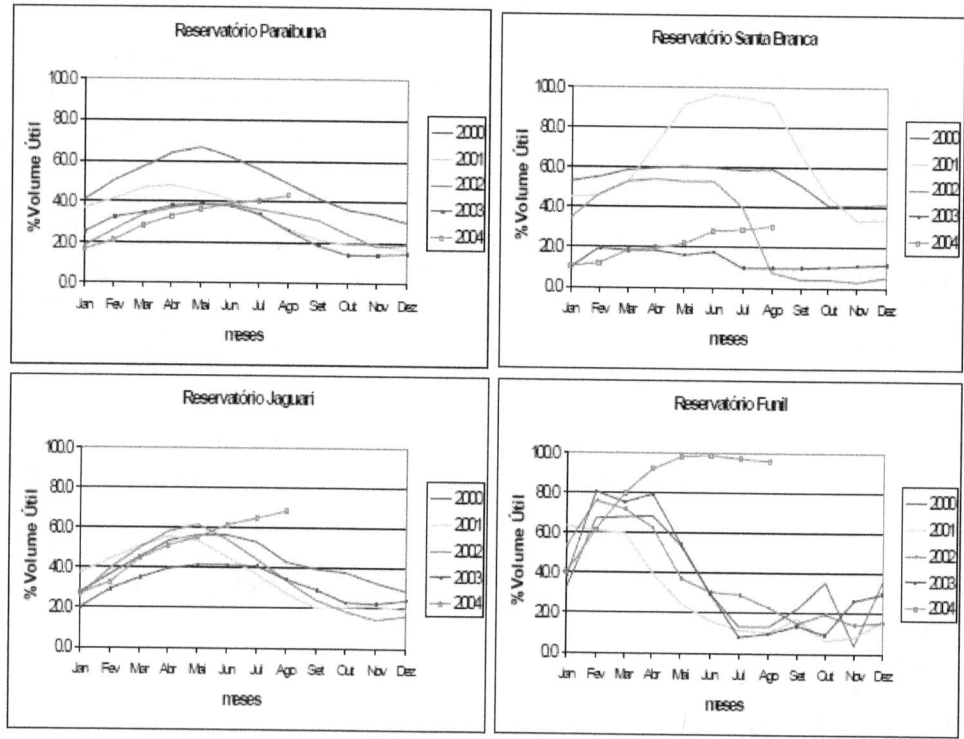

Figure 12. Useful volumes of the Paraibuna, Santa Branca, Jaguari and Funil reservoirs (2000-2004 period)

Figure 12 shows the useful volumes of the Paraibuna, Santa Branca, Jaguari and Funil reservoirs, for the period from January 2000 to August 2004, showing a significant recovery of the reservoirs after the application of the operating rules defined by the new resolution.

6. CONCLUSIONS

The natural flows affluent to the Paraibuna and Jaguari reservoirs for the period from 1996 to 2001, were, in fact, lower than the average long-term flows. Regarding the precipitation series, it was found that the decrease in average precipitation, presented in the basin, was not sufficient to explain the current low level of accumulation in the basin's reservoirs.

Considering that the basin's reservoirs have not been able to recover their useful volumes in recent years and considering the importance of the Paraíba do Sul basin for the supply of several cities, including part of the Metropolitan Region of Rio de Janeiro, a Resolution on the rules to be issued to be adopted, for the operation of the hydraulic system of the Paraíba do Sul River, comprising in addition to the reservoirs located in the basin, also the structures for transposing the waters of the Paraíba do Sul River to the Guandu system.

The new operating rules, defined with the aid of the AcquaNet flow network model, proved to be efficient in recovering the reservoirs in the Paraíba do Sul River hydrographic basin.

7. BIBLIOGRAPHIC REFERENCES

Fonte: (http://www.cptec.inpe.br/%7energia)

LABADIE, J. W. - Program Modsim: River Basin Network Flow Model For The Microcomputer, Department of Civil Engineering, Colorado State University, 1988.

Resolução ANA Nº211, de 26 de maio de 2003.

SIPOT, 2003 – Sistema de Informações do Potencial Hidrelétrico Brasileiro, Versão 4.0, ELETROBRÁS.

CHAPTER 9

Drought Monitoring in the Northeast of Brazil

Marcos Airton de Sousa Freitas
Senior Water Resources Specialist with National Water Agency

ABSTRACT: It is extremely important to have instrumental aid to decision making to reduce disaster risks related to extreme hydrometeorological events such as drought. In this sense, several indexes have been applied to several States of the Northeast Brazil, so that different mitigating actions could be implemented in accordance with values achieved by these parameters. The following meteorological indices were used: the RAI (Rainfall Anomaly Index), the BMDI (Bhalme & Mooley Drought Index) and LRDI (Lamb Rainfall Departure Index). A practical advantage in using these indices is the monthly monitoring of the degree of severity and duration of dry periods, to mitigate the impacts caused by these extreme events.

KEYWORDS: drought index, drought, Northeast Brazil

1. INTRODUCTION

Precipitation and its extreme values (floods and droughts) in northeastern Brazil are partly dependent on global climatic phenomena, such as, for example, the Inter-Tropical Convergence Zone (ZCTI), El Niño, the South Oscillation, Dipole of the Atlantic and others (Freitas, 2010). Droughts are clearly different from other natural disasters. Contrary to other natural occurrences such as floods, hurricanes and earthquakes, which start and end suddenly, in addition to being normally restricted to a small region, the phenomenon of droughts has, almost always, a slow start, a long duration and spread if, for the most part, over an extensive area.

According to Martins et al. (2013), the present climatic variability of the Northeast Region already imposes great challenges to the management of water resources. However, the projections of climate changes, in turn, indicate that this situation may worsen for the region. The use of this information for the planning of the sector is still a challenge, either due to the lack of theoretical tools not yet fully developed, or due to the need for specific studies for the basin or region of interest.

In the Management of Water Resources in semi-arid regions, such as in the Northeast of Brazil, it is extremely important to have practical tools to assist decision-making, especially during periods of drought. Freitas (1996) proposed the implementation of a Decision Support System for monitoring droughts for the Brazilian semiarid, considering the meteorological indexes. Several indexes have been adapted and incorporated into a system for monitoring the basic characteristics of drought periods, namely, duration, severity and intensity, so that different mitigating actions could in fact be implemented, according to the values reached by these parameters. In the monitoring of drought in the region, the following indices were used: RAI (Rainfall Anomaly Index), BMDI (Bhalme & Mooley Drought Index), as well as LRDI (Lamb Rainfall Departure Index).

1. DROUGHT RISK MANAGEMENT

Among the objectives of the National Water Resources Policy - PNRH, instituted by Law nº 9.433, of January 8, 1997, is found in its Art. 2, § III, the prevention and defense against critical hydrological events of natural origin or arising from the inappropriate use of natural resources. In Law No. 9,984, of July 17, 2000, which provides for the creation of the National Water Agency - ANA, the federal entity for the implementation of the National Water Resources Policy and for the coordination of the National Water Resources Management System, confirm at its Art. 4 that, ANA's performance will obey the foundations, objectives, guidelines and instruments of this National Policy, being responsible, among other activities, for planning and promoting actions aimed at preventing or minimizing the effects of droughts and floods, in the within the National Water Resources Management System, in conjunction with the central body of the National Civil Defense System, in support of States and Municipalities.

Decree No. 7,257, of August 4, 2010, which aims to regulate Provisional Measure No. 494 of July 2, 2010, provides for the National Civil Defense System - SINDEC, on the recognition of emergency situations and the state of emergency. public calamity. In its Art. 5 § 6, it establishes that "to coordinate and integrate the actions of SINDEC throughout the national territory, the National Secretariat for Civil Defense will maintain a national center for risk and disaster management, with the purpose of streamlining the actions of response, monitor the most prevalent disasters, risks and threats ".

According to UNISDR (2004), disaster is defined as "an event that may cause physical damage, a phenomenon or human activity that may cause loss of life or injury, damage to property, social and economic disruption or environmental degradation". Disasters can include latent conditions, which can represent future threats and can have different origins: natural (geological, hydrometeorological and biological) or induced by human processes (environmental degradation and technological disasters). Disaster Risk Reduction (DRR) includes all policies, strategies and measures that can make people, towns, cities and countries more resistant to risks and reduce risk and vulnerability to disasters, namely: prevention; mitigation; preparation; recovery and reconstruction (UNISDR, 2004).

Among these activities, it is worth mentioning drought monitoring, with the following uncertainty in this process: the identification of drought rates; collection, processing and transmission of data and operational uncertainties. The risks included: inadequate monitoring parameters (indexes); lack of financial resources; identify and improve appropriate indexes for the region; improve information systems and databases; elaborate vulnerability and risk maps (Freitas, 2010). The aim of this article is to address some of these deficiencies.

3. DROUGHT INDEXES

In drought monitoring, indices are normally used as a measure of the severity of a dry period. According to their formulation, the indexes can be classified into meteorological, hydrological and agricultural. Although precipitation is an important factor, the climate in given region should not be classified as dry or wet based on precipitation series alone. Evapotranspiration plays, particularly in semi-arid regions like Northeast Brazil, a key role. In addition, precipitation and evapotranspiration come from different meteorological causes. It is necessary to observe whether the precipitation is greater or less than the evapotranspiration, among several other aspects. Below, several indices are investigated to determine the possibility of their practical applications in a drought monitoring system in Northeast Brazil.

Rainfall Anomaly Index (RAI)

Freitas (1999) implemented and applied the Rainfall Anomaly Index (RAI) pioneering in Brazil, incorporating it into a drought management monitoring system (Freitas, 1998; Freitas, 2005). To make the deviation of precipitation in relation to the normal condition of several regions subject to comparison ROOY (1965) presented the following index:

$$RAI = 3.\left[\frac{(N - \bar{N})}{(\bar{M} - \bar{N})}\right]$$, for positive anomalies

$$RAI = -3.\left[\frac{(N - \bar{N})}{(\bar{X} - \bar{N})}\right]$$, for negative anomalies

Where: $N =$ current monthly precipitation; $N =$ average monthly rainfall of the historical series; $M =$ average of the ten largest monthly rainfall in the historical series; and $X =$ average of the ten lowest monthly rainfall in the historical series.

Figure 1 shows the application of this index for two rainfall stations in the State of Rio Grande do Norte. Based on this index, it is possible to make a comparison of current precipitation conditions with historical values. It also serves to assess the spatial distribution of a drought, depending on its intensity.

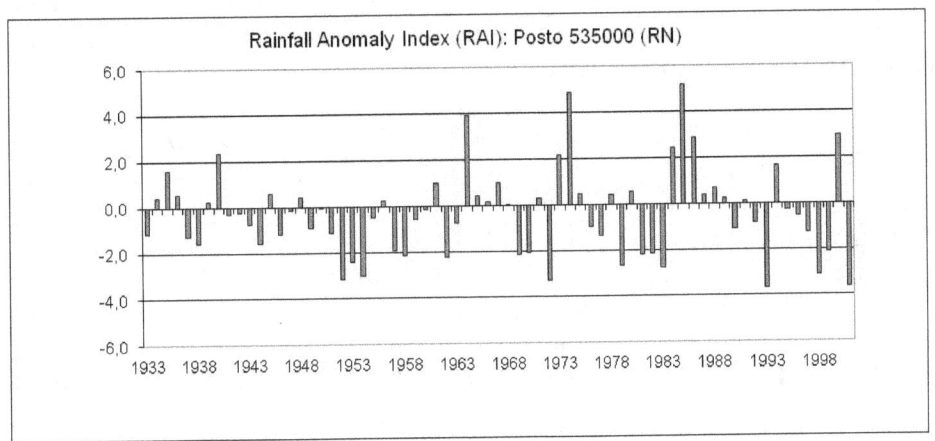

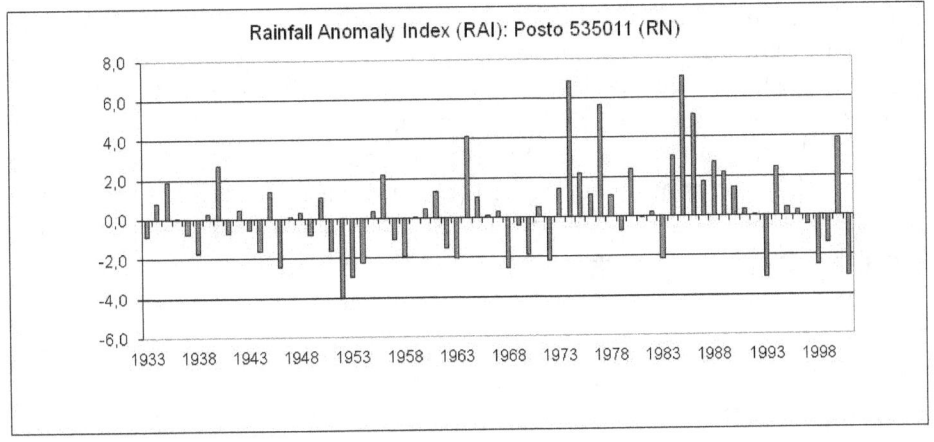

Figure 1: Rainfall Anomaly Index (RAI) for two rainfall stations in the State of Rio Grande do Norte.

Bhalme & Mooley Drought Index (BMDI)

PALMER (1965) presented a water balance procedure, which was later known as the Palmer Drought Severity Index (PDSI), for the semi-arid region of western Kansas and for the sub-humid region of Iowa, in the United States. PDSI is calculated based on evapotranspiration, infiltration, eventual runoff data, etc. and expresses a measure for the accumulated difference between normal precipitation and the precipitation required for evapotranspiration. This analysis is done on a weekly or monthly level. This procedure results in an index that varies from -4 (extreme droughts), passing through zero (normal conditions) to +4 (very humid periods).

ALLEY (1984) and GUTTMAN (1991) demonstrated that the PDSI was not a good indicator of humidity conditions, particularly in dry periods. Another disadvantage of the PDSI results from the fact that the regularization of the surface flow is not considered. In a study in tropical regions of India, BHALME & MOOLEY (1979, 1980) highlighted these problems. They then proposed a modification of the original index to incorporate the climatic conditions in force in India. Such an index became known as the Bhalme & Mooley Drought Index (BMDI). The application of this new index for the State of Rio Grande do Norte is presented here.

As it presents positive and negative values, this index can be used to evaluate periods of drought and floods. The average value for the 1st semester of the year, applied to stations in Rio Grande do Norte (1931-2000) is shown in figure 2, to those in the State of Paraíba (Figure 3), Piauí (Figure 4), Pernambuco (Figure 5) and Sergipe (Figure 6). The current, monthly, accumulated value of BMDI during the crop growth period or the rainy season (January to June) can then be compared with the historical values of the region, in order to have a permanent control of the humidity condition.

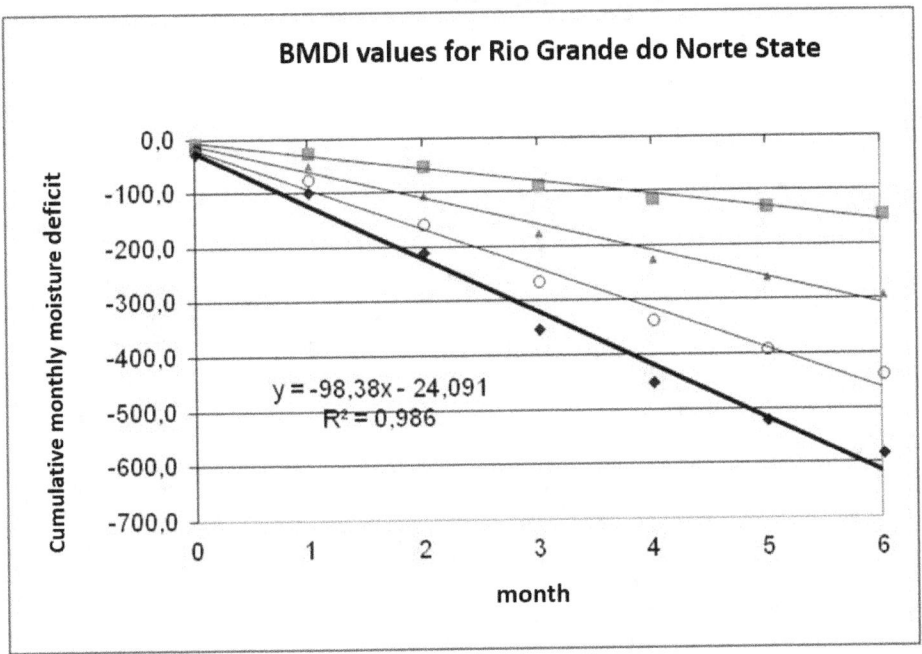

Figure 2: BMDI values for Rio Grande do Norte State

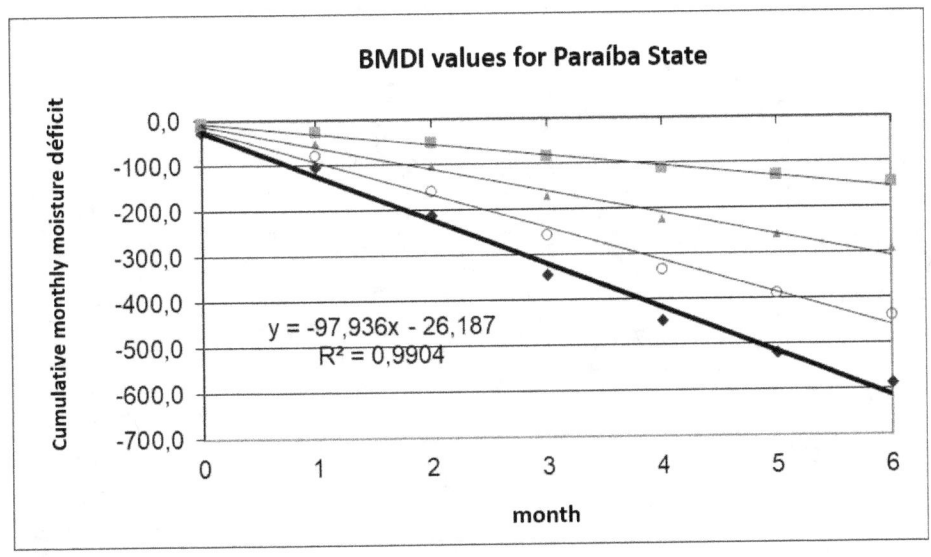

Figure 3: BMDI values for Paraíba State

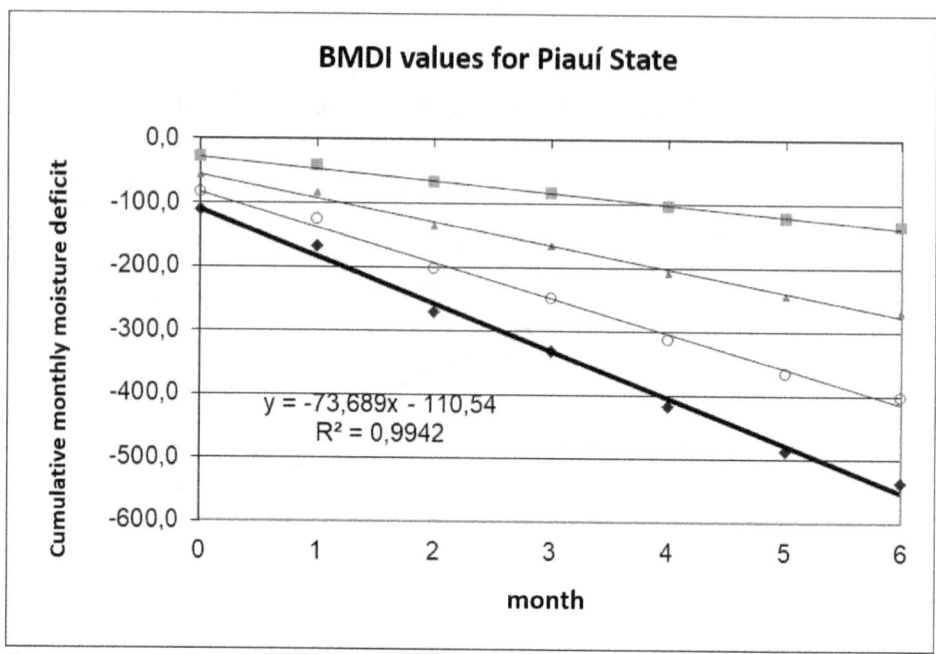

Figure 4: BMDI values for Piauí State

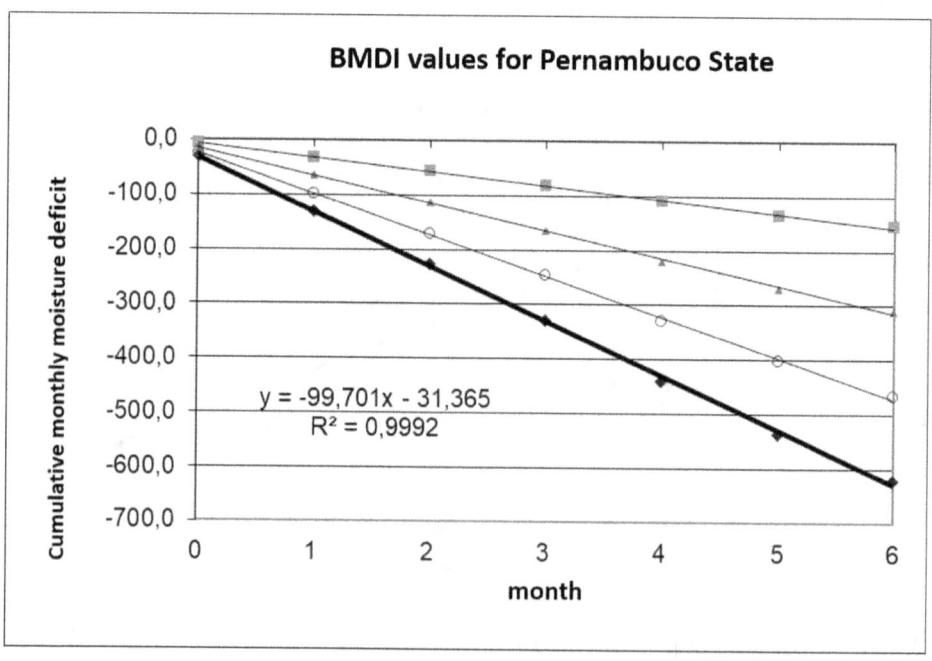

Figure 5: BMDI values for Pernambuco State

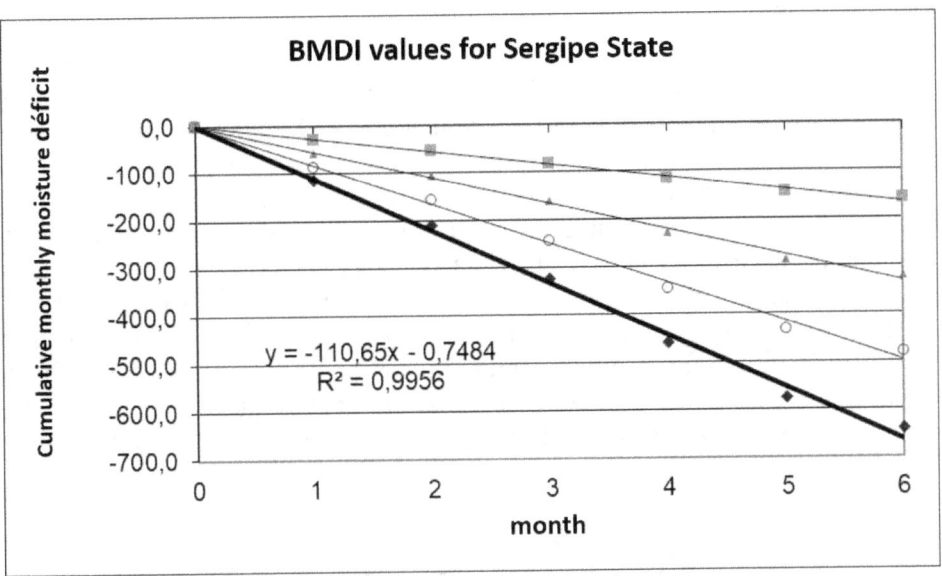

Figure 6: BMDI values for the State of Sergipe

Lamb Rainfall Departure Index (LRDI):

The calculation of this index (LAMB et al., 1986) consists of a normalization procedure, through which the average deviations of precipitation of several stations in given region are grouped in the determination of a single index, given by:

$$LRDI = \frac{1}{t_j} \cdot \sum_{i=1}^{t_j} \frac{N_{i,j} - \overline{N}_i}{S_i}$$

where $N_{ij} =$ precipitation in year j of station i; Ni = average annual precipitation of station i; Si = standard deviation of the annual precipitation of station i; e tj= number of stations with precipitation in year j.

A major advantage of this method is that all precipitation series, which normally have many flaws, can thus be used to determine a regional index. Figure 7 presents the result of the precipitation of this methodology to the analyzed rainfall stations, in the period 1931-2000, in the State of Rio Grande do Norte.

4. CONCLUSIONS

Several meteorological indexes - RAI (Rainfall Anomaly Index), BMDI (Bhalme & Mooley Drought Index) and LRDI (Lamb Rainfall Departure Index) - have been modified and incorporated into a Decision Support System (SSD) to monitor the basic characteristics of the periods drought, that is, duration, severity and intensity, so that different mitigating actions could actually be implemented, according to the values reached by these parameters. A crucial advantage in the use of these indices is the almost simultaneous monitoring of the degree of severity and duration of dry periods, allowing effective measures to be taken in a timely manner, aiming to lessen the impacts caused by a drought. It is recommended to apply this methodology to all rainfall stations in the semiarid region of Brazil, resulting in a better spatial distribution of these values.

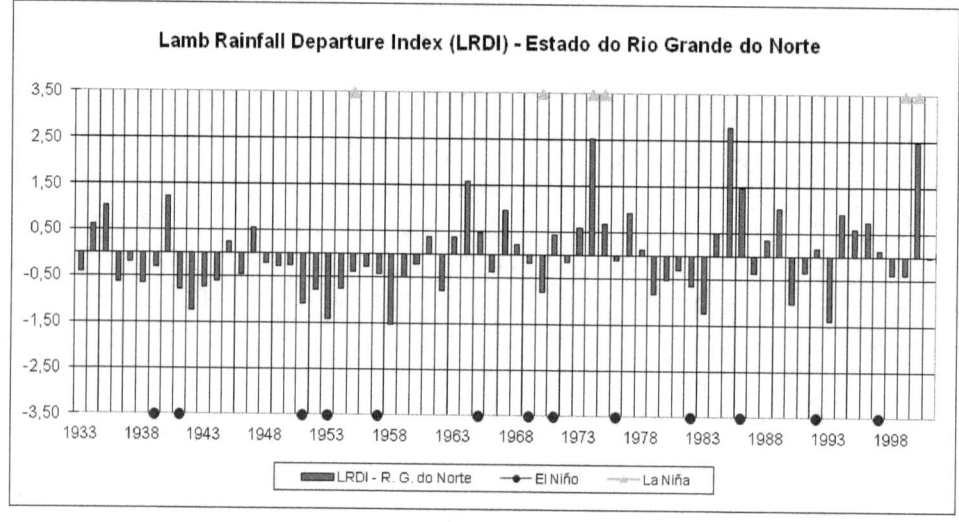

Figure 7: LRDI values for the state of Rio Grande do Norte.

5. BIBLIOGRAPHIC REFERENCES

ALLEY, W.M., 1984: **The Palmer Drought Severity Index: Limitations and Assumptions**, Journal of Climate and Applied Meteorology, 23, 1100-1109.
ARAÚJO, Lincoln Eloi, Aline Costa Ferreira, João Miguel de Moraes Neto, Francisco de Assis Salviano de Sousa. **VARIABILIDADE ESPAÇO-TEMPORAL**

DA PRECIPITAÇÃO NO CARIRI PARAIBANO Revista Educação Agrícola Superior - v.22, n.2, p.23-26, 2007.

BHALME, H.N., D.A. MOOLEY, 1979: **On the Performance of Modified Palmer Index**, Archives for Meteorology, Geophysics, and Bioclimatology, Ser. B, 27, 281-295.

BHALME, H.N., D.A. MOOLEY, 1980: **Large-Scale Drought/Floods and Monsoon Circulation**, Monthly Weather Review, 108, 1197-1211.

CHAGNON, S.A., 1980: **Removing the Confusion over Droughts and Floods: The Interface between Scientists and Policy Makers**, Water International, 10-18.

FREITAS, M.A.S, 2010: **Que Venha a Seca: modelos para gestão de recursos hídricos em regiões semiáridas**, Ed. CBJE, Rio de Janeiro, 413p.

FREITAS, M.A.S. & M.H.A. BILLIB, 1997: **Drought Prediction and Characteristic Analysis in Semi-Arid Ceará / Northeast Brazil**, Symposium "Sustainability of Water Resources Under Increasing Uncertainty", IAHS Publ. No. 240, 105-112, Rabat, Marrocos.

GUTTMAN, N.B., 1991: **A Sensitivity Analysis of the Palmer Hydrologic Drought Index**, Water Resources Bulletin, 27(5), 797-807.

GUTTMAN, N.B., J.R. WALLIS, & J.R.M. HOSKING, 1992: **Spatial Comparability of the Palmer Drought Severity Index**, Water Resources Bulletin, 28(6), 1111-1119.

LAMB, P.J., R. A. PEPPLER & S. HASTENRATH, 1986: **Interannual Variability in the Atlantic**, Nature, 322, 238-240.

Martins, E.S.; Braga, C.F.C.; Souza Fo., Francisco Assis de; Moraes, Marcia M.G.A. ; Marques, G. F. ; Mediondo, E.M. ; Freitas, Marcos Airton de Sousa ; Vazquez, V. ; Engle, Nathan ; Denys, E. . **Climate change impacts on water resources management: adaptation challenges and opportunities in Northeast Brazil**. In: The World Bank Group. (Org.). Latin America and Caribbean Region Environment and Water Resources occasional paper series. 1ª ed., Washington DC: World Bank, 2013, v. 1, p. 1-6.

McDONALD, N.S., 1989: **Decision Making using a Drought Severity Index**, Proc. United Nations University Workshop, Need for Climate and Hydrologic Data in Agriculture in Southeast Asia, CSIRO Division of Water Research, Technical Memo 89/5.

PALMER, W.C., 1965: **Meteorological Drought**, Weather Bureau, U.S. Department of Commerce, Washighton, D.C., Research Paper n° 45, 1-58.

ROOY, M.P. van, 1965: **A Rainfall Anomaly Index Independent of Time and Space**, Notos, 14, 43.

United Nations International Strategy for Disaster Reduction Secretariat (UNISDR) - **Living with risk: a global review of disaster reduction initiatives**, Geneva, 2004, 429p.

CHAPTER 10

The Piancó-Piranhas-Açu Hydrographic Basin Face to the 2012-2020 Drought Event

Marcos Airton de Sousa Freitas

Specialist in Water Resources at the National Water Agency - ANA; Ministry of Regional Development - MDR; masfreitas@ana.gov.br

Abstract: The semiarid region of Northeast Brazil experienced during the years 2012 to 2020 one of its worst periods of severe drought. This was largely due to the conditions of the El Niño phenomenon in the Pacific Ocean and the oceanographic and atmospheric conditions of the Atlantic Ocean. In this article, it was intended to address the aspects of water management due to this extreme drought, both from the supply and demand side, in a typical hydrographic basin in the Brazilian semiarid, the Piancó-Piranhas-Açu river basin. With the reduction of rainfall and the consequent tax flow to the main reservoirs in the region, the operation of the reservoirs responsible for the main uses of the region's water resources is shown to be of fundamental importance: irrigated agriculture (with an estimated area of around 50 hectares in the basin) ; aquaculture; human and industrial supply, among others. Actions aimed at the rational use of water, monitoring and negotiated water allocation mechanisms, as well as communication were also carried out.

Keyword: water management, water demand, drought

Resumo: A região semiárida do Nordeste do Brasil experimentou durante os anos de 2012 a 2020 um de seus piores períodos de seca severa. Isso se deveu, em grande parte, às condições do fenômeno El Niño, no Oceano Pacífico, e às condições oceanográficas e atmosféricas do Oceano Atlântico. Neste artigo, pretendeu-se abordar os aspectos da gestão dos hídricos em função dessa seca extrema, tanto pelo lado da oferta quanto pela demanda, em uma típica bacia hidrográfica do semiárido brasileiro, a bacia dos rios Piancó-Piranhas-Açu. Com a redução das chuvas e consequente vazão tributária aos principais reservatórios da região, mostra-se como de fundamental importância a operação dos reservatórios responsáveis pelos principais usos dos recursos hídricos da região: agricultura irrigada (com área estimada em cerca de 50 hectares na bacia); aquicultura; abastecimento humano e industrial, entre outros. Ações voltadas para o uso racional da água, monitoramento e mecanismos de alocação negociada de águas, bem como de comunicação também foram realizadas.

Palavras-chave: gestão de recursos hídricos, demanda de água, seca

1. Introduction

The Piranhas-Açu river runs through the states of Paraíba and Rio Grande do Norte, reaching the Atlantic Ocean in the vicinity of the city of Macau, thus being a river under the Union's domain. It has two springs depending on the criterion to be adopted. Geographically, it is placed in the municipality of Bonito de Santa Fé, on the border between Paraíba and Ceará. When the criterion of the largest drainage area is adopted, according to ANA Resolution No. 399/2004, the source is located in the Serra de Piancó and the Piancó River becomes the main water body, hence the hydrographic basin of the Piancó-Piranhas-Açu (PRH Piranhas-Açu – Diagnóstico, 2014).

The Piranhas River, in the State of Paraíba, forms a hydrographic system consisting of the Peixe and Piancó River basins and part of the Espinharas and Seridó River basins. These 4 (four) rivers are its main tributaries - the first, on the left bank, and the last three, on the right bank. Still in Paraíba, it receives contributions from smaller water courses, such as the Campos, Cachoeira, Corda and Trapiá streams, on its right bank, and Paraguay, Solidão and Tamanduá streams, on the left bank. In Rio Grande do Norte, the Piranhas River enters the municipality of Jardim de Piranhas, receives the waters of the Espinharas and Seridó rivers and crosses the central region of the State. After the Armando Ribeiro Gonçalves dam, which together with the Curema-Mãe D´água System represents the large reservoirs of superficial water storage in the basin, the Piranhas River changes its name to Piranhas-Açu.

It is the responsibility of the National Water Agency (ANA), according to Law 9.984 / 2000, "to define and inspect the conditions of operation of reservoirs by public and private agents, aiming to guarantee the multiple use of water resources, as established in the water resource plans respective river basins ". The responsibility to "plan and promote actions aimed at preventing and minimizing the effects of droughts and floods, within the scope of the National Water Resources Management System, in conjunction with the central body of the National Civil Defense System, in support of States and Municipalities ", Is also attributed to ANA by this law.

The aim of this article is to address the aspects of water resource management, both from the supply side and from the demand side, in view of the extreme drought, covering the period from 2012 to 2020.

2. Methodology

ANA Resolution No. 687/2004, provided for the Regulatory Framework for the management of the so-called Curema-Açu System and established parameters and conditions for issuing preventive grants and the right to use water resources. The Curema-Açu System, for the purpose of that Resolution, was divided into six sections (Figure 1). According to Lima et al. (2000), in this system is the only hydroelectric plant in the State of Paraíba, the hydroelectric plant of the Curema reservoir, responsible for the generation of 3.52 MW of energy. After passing through the Curema turbines, the waters flow into the Piancó River, perpetuating the Piranhas-Açu River, reaching the State of Rio Grande do Norte. Part of this water is used to meet the system's downstream demands: Coremas / Sabugi pipeline, downstream irrigation, including the flow required by Rio Grande do Norte, established in the Regulatory Framework, in addition to the São Bento do Brejo do Cruz pipelines and pipeline scored by Catolé do Rocha. In the Mãe D'água reservoir, the 37 km long Canal da Redenção starts, which aims to supply the water demand of the Várzeas de Sousa irrigation project (5,000 ha), in the Rio do Peixe basin.

According to Art. 12 of ANA Resolution 687/2004, the generation of energy from the hydroelectric power plant of the Curema reservoir, owned by the Companhia Hidrelétrica do São Francisco - CHESF, cannot compromise the maximum available flows established, notably in Sections 2 (Rio Piancó), 3 (Rio Piranhas - PB) and 4 (Rio Piranhas - RN).

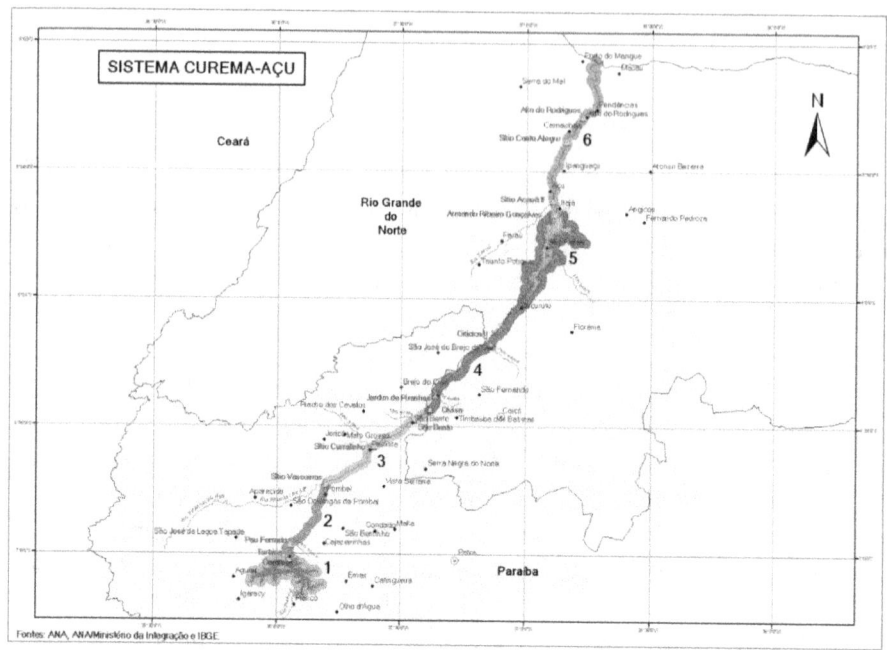

Figure 1: Piancó-Piranhas-Açu river basin

From the evolution of the volume of the equivalent reservoir (considering reservoirs with a capacity above 10 hm³), it appears that the State of Paraíba was in September 2012, with 54.1% of the volume of the equivalent reservoir. This figure dropped to 35.2%, in the same month in 2013, and to 28% in 2014, reaching 18.9% in September 2015. In May 2016, this figure was in the order of 16.4%. Rio Grande do Norte was in September 2012, with 61.6% of the volume of the equivalent reservoir. This figure dropped to 42.3%, in the same month in 2013, and to 35.7% in 2014, reaching 25.2% in September 2015. In May 2016, this figure was in the order of 21, 8% (ANA, 2016). The Northeast, in general, has been going through the year of 2012 going through a period of extreme drought, which continues in 2016 (Figure 2).

Depending on the PRH Piranhas-Açu - Diagnosis (2014), it is worth mentioning that, in the context of the semiarid climate, the rivers that form the hydrographic basin are intermittent. The storage of water to serve different uses is ensured by several reservoirs, of a strategic character (accumulation capacity close to or greater than 10 hm³), which together reach more than 5,000 hm³.

In this sense, it is of huge importance the development of models that can be used for drought prediction, such as those developed by Freitas (1996), Freitas and Billib (1997) and also Freitas and Freitas (2020), as well as the use of models of synthetic streamflow generation for intermittent rivers (Freitas, 1995; Freitas et al., 2019, Freitas e Freitas, 2019a, Freitas e Freitas, 2019b, Freitas e Freitas, 2019c). Based on these models, it is possible to generate scenarios for different types of droughts, associated with their main characteristics (duration, severity and magnitude).

The Armando Ribeiro Gonçalves reservoir (with 541.94 hm^3, representing 22.6% of the maximum capacity, on 05/31/2016), in Rio Grande do Norte, the Curema reservoir (with 48.51 hm^3, representing 8.2% of the maximum capacity, on 05/31/2016), Mãe D'Água reservoir (with 76.42 hm^3, representing 13.5% of the maximum capacity, on 05/31/2016) and Avid Engineer reservoir (with 20.03, representing 7.9% of the total capacity , on May 31, 2016), in Paraíba, correspond to about 70% of the basin's storage capacity. These reservoirs are responsible for the perpetuation of stretches of river downstream where different uses of water are developed.

The water supply is represented not only by these superficial reservoirs, but also by a great number of wells and rainwater catchment systems (cisterns) in the hydrographic basin. The dominance of rivers and reservoirs in the hydrographic basin forms a mixture of water bodies under the responsibility of both the States and the Union, which demonstrates the importance of the integrated water resources management in the hydrographic basin. On the water demand side, the development of the agricultural sector, driven in large part by the growing demand for food, generates demand for water to supply the natural needs of crops. With this, this sector is configured as an important user of water resources, whose behavior must be quantified to planning and management actions to be effective (Freitas and Lopes, 2003).

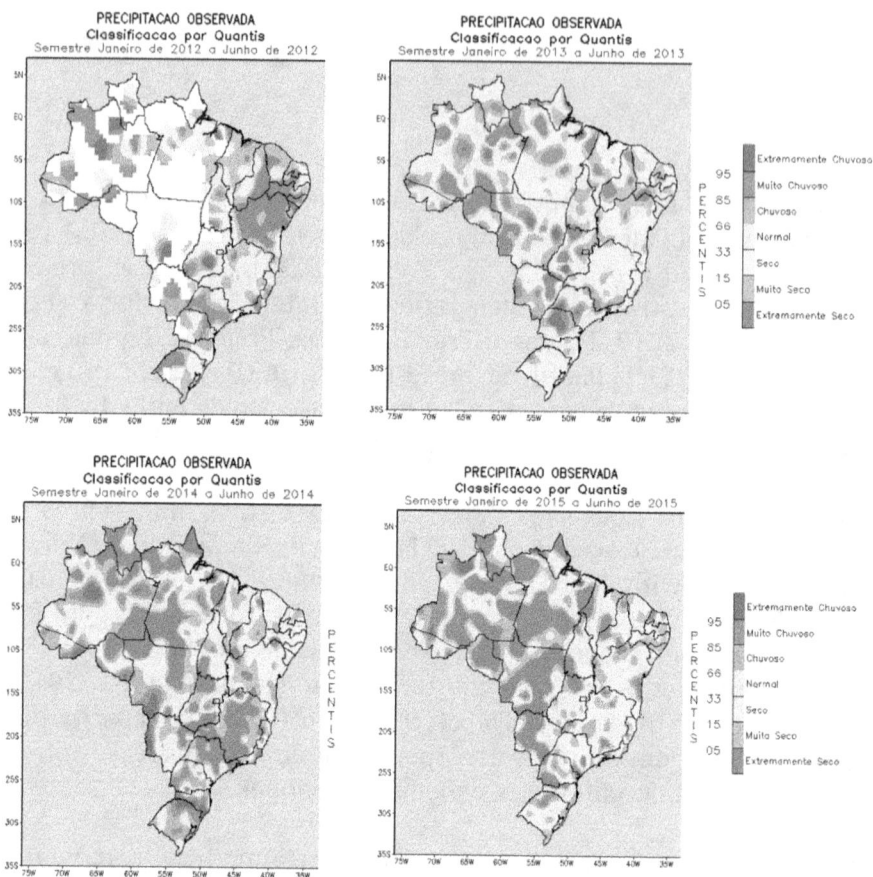

Figure 2: Observed precipitation (Classification by Quantiles), for the period 2012-2015 (Source: National Institute of Meteorology - INMET).

Irrigation consists of a set of techniques aimed at displacing water in time and space, modifying the agricultural possibilities of each region, in order to correct or adjust the natural distribution of availability to the needs of the crops. The crop water needs vary depending on the growth stage of the plants, the type of soil and the climatic conditions of the region and are supplied by water from precipitation, complemented by irrigation. In general, irrigation is in conflict to other water users, such as electricity generation and human consumption (Freitas, 2009; Freitas, 2010).

According to the PRH Piranhas-Açu, the water demand (withdrawal flow) to meet the various uses in the basin was of the order of 41.4 m³/s, while the consumption flow totaled 23.8 m³/s. The irrigation activity represents a demand of 27.3 m³ s, representing 65.7% of the total consumption demand in the basin. There are two irrigation models in the basin: irrigation concentrated in perimeters sponsored by the Federal or State governments, and diffuse irrigation, represented by small private areas, in small and medium-sized properties, especially along the perennial stretches and around the hydraulic basin of reservoirs.

3. The Curema-Mãe D´água and Armando Ribeiro Gonçalves Reservoirs System

In the case of the hydrographic basin, the Curema-Mãe D'água System stands out, which supplies 250 thousand people in the states of Rio Grande do Norte and Paraíba. In both, measures were adopted to restrict irrigation demand, even limiting days and times for this practice. For complying with these restrictions, ANA has been carrying out regular inspection campaigns. Bathymetric surveys were also carried out in strategic reservoirs to verify the real storage capacity, as well as numerous flow measurement and inspection campaigns. All these actions allowed a better estimate, both stored volume and the flow released to meet the multiple uses downstream of the reservoir.

Figure 3: Mãe D´água and Curema Reservoirs.

Figures 5, 6 and 7 show, respectively, the evolution of the volume of the Curema, Mãe D'água and Armando Ribeiro Gonçalves reservoirs, due to the restrictions of uses implemented in the basin and the flows released by the reservoir in the years 2013 to 2020. It appears that for the year 2013 the water accumulated volume (net volume) of the Curema reservoir in the rainy season was around 50 hm^3. In 2014, with the actions of control, management and inspection, water accumulated volume (net volume) was 100 hm^3. In 2015, in this value dropped to 25 hm^3 and in 2016, it was only around 5 hm^3. For the year 2013, the water accumulated volume (net volume) in the rainy season in the Mãe D'água reservoir was 10 hm^3. In 2014, this net volume was 35 hm^3. In 2015, this value dropped to 20 hm^3 and in 2016, it was only 6 hm^3.

Figure 4: Mãe D'água Reservoir and Redenção water channel for irrigation.

For the year 2013, it appears that the water accumulated volume (net volume) in the rainy season in the Armando Ribeiro Gonçalves reservoir was of the order of 40 hm^3. In 2014, this net volume was 300 hm^3. In 2015, this value dropped to 40 hm^3 and in 2016, it was around 90 hm^3.

Taking into account the possibility of the large reservoirs being compromised, which would damage the supply systems of the main cities in the Northeast, if 2014 was also a year with below average rains, ANA started to adopt measures in order to adapt supply and demand to avoid shortages of the population.

Then, there was a reduction in the released streamflow from the reservoirs (Figure 8) and the fixing of alternate days for the water abstraction in rivers and dams for productive activities. The average released streamflow in the period from June to March (2013/14, 2014/2015 and 2015/2016), were, respectively, 4.19 m³/s; 5.88 m³/s and 1.3 m³/s.

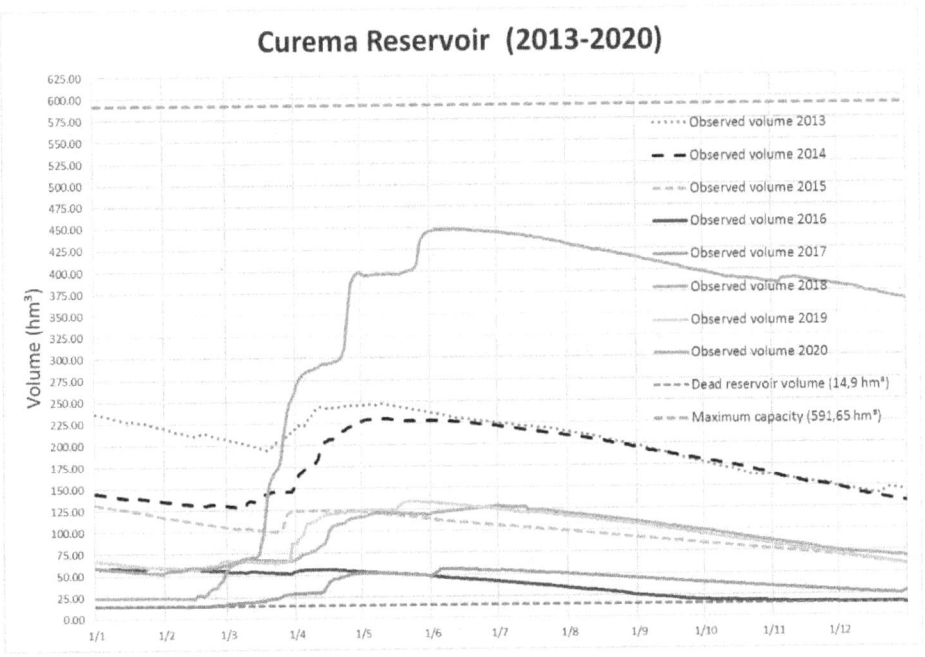

Figure 5: Evolution of the Curema Reservoir Volume

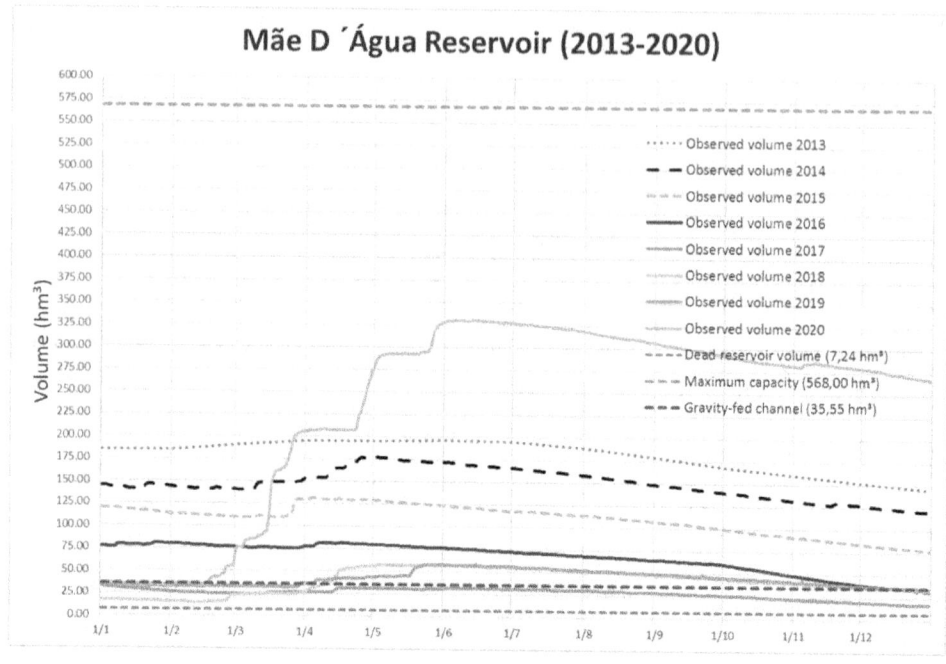

Figure 6: Evolution of the Mãe D´Água Reservoir Volume

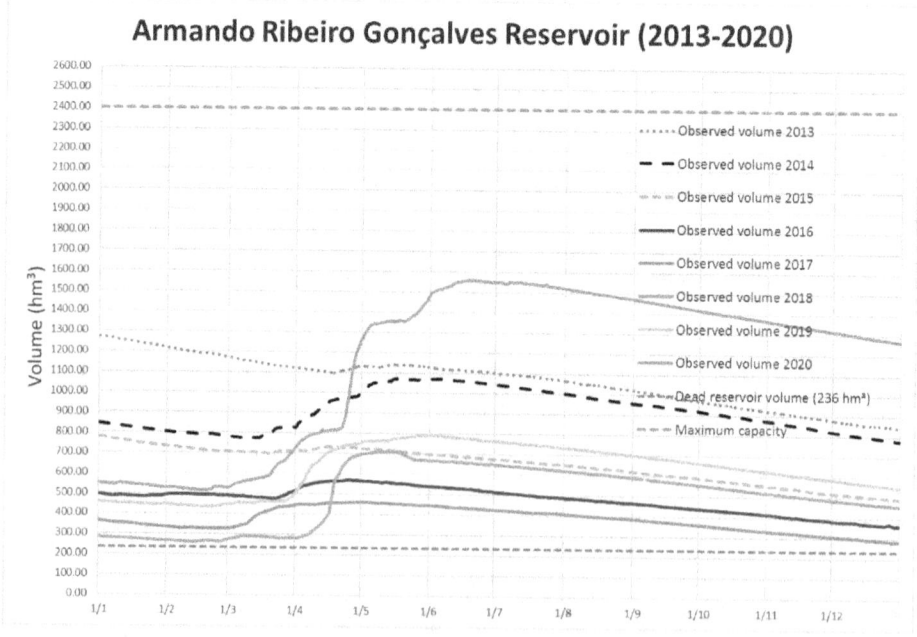

Figure 7: Evolution of the Armando Ribeiro Gonçalves Reservoir Volume

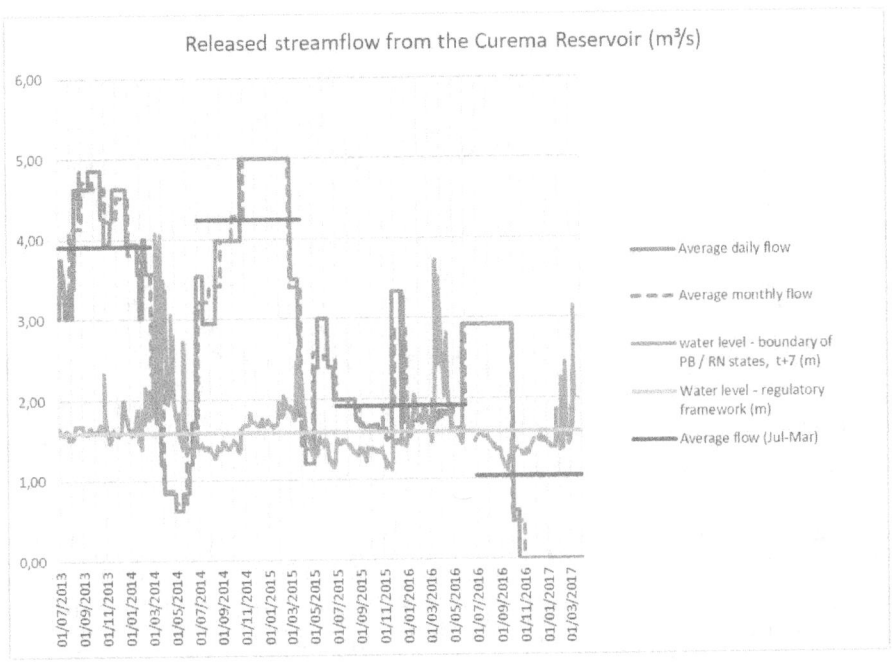

Figure 8: Released streamflow from the Curema Reservoir (m³/s)

It is worth mentioning that the flow regularized with a 100% guarantee for the Curema reservoir, according to the Report of the Ministry of Integration (MI, 2000) is around 8 m³/s. Initially, rules for restricting use were established on the Piranhas-Açu river and in the Curema-Mãe D'água reservoirs system, Ávidos, São Gonçalo, Itans and Santa Inês reservoirs. Later, with the worsening of the crisis, the suspension of uses for irrigation and aquaculture was declared (Figure 9).

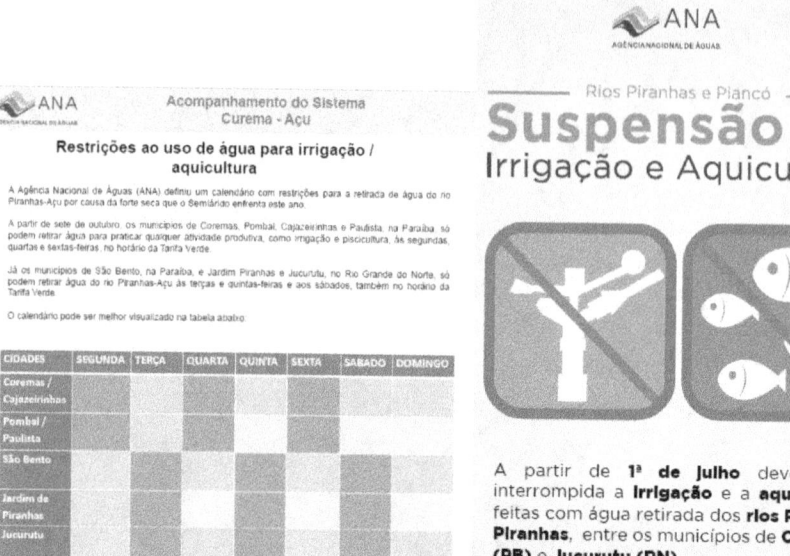

Figure 9: Disclosure posters for the Restriction of Uses and the Suspension of Uses for Irrigation and Aquaculture

Water deficits as result of non-sustainable water resources development, are the main reason for water related conflicts in water scarcity affected regions, especially due to strong competition between the different water users. According to Rusteberg and Freitas (2018), new approaches, methods and tools have been developed to support the integrated planning and management of water resources towards the sustainable development of the region. Non-conventional water resources, such as treated wastewater, brackish or imported water needs to be part of an integrated strategy for IWRM implementation.

According to Lopes & Freitas (2007) and Freitas (2009), water allocation in Brazil, historically, is characterized by a strong intervention of the public sector. Water allocation can be understood as an Integrated Water Resources Management (IWRM) measure that aims to provide water to current and future users of the water resources system, matching water supply and demand, even meeting environmental demands, being in line with strategic management objectives. In this sense, there are several mechanisms of water allocation, which operate from public authority guidelines, negotiation processes among water users or from technical concepts, such as the limits of the use of water bodies, or economic, such as charging for water use (Rusteberg and Freitas, 2018).

For the so-called Negotiated Water Allocation a methodology is described in Freitas, M.A.S. (2003); ANA/GEF/PNUMA/OEA (2004); Freitas, M.A.S (2010) and improved by Freitas (2013), Martins et al., (2013). This methodology was largely internalized in the administrative procedures of the National Water Agency, through Technical Note No. 10/2015/COMAR/SRE (2015) and has been applied throughout the semi-arid region of North-East Brazil.

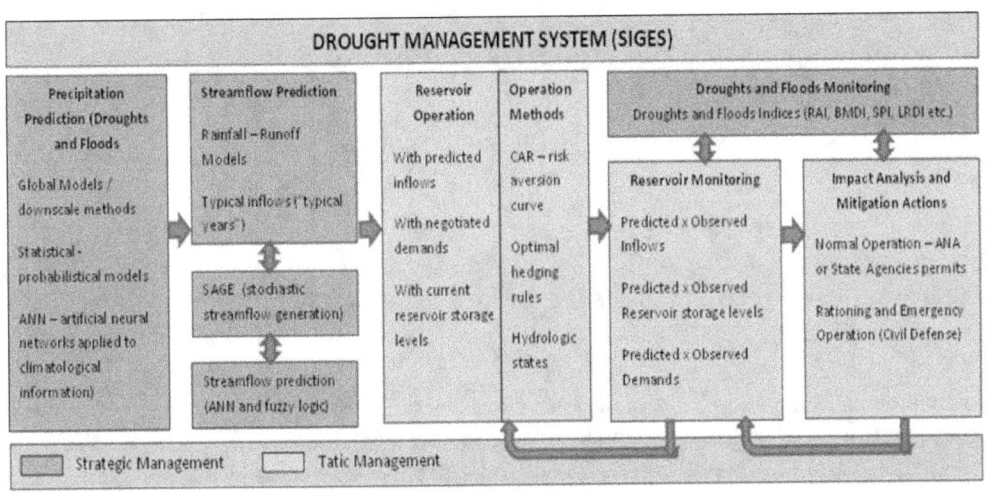

Figure 10: Drought Management System - SIGES (Freitas, 2013)

Figures 11 and 12 show the *Risk Aversion Curves* - CAR proposed, respectively, for the Curema-Mãe D'água system and for the Armando Ribeiro Gonçalves reservoir (Freitas and Gondim Filho, 2007).

The operation of reservoirs based on the *Risk Aversion Curves* is guided by a probability distribution of flows flowing into the reservoir system and is useful for determining the water allocation among users with different guarantees. The figures show the useful reservoir volume (%) on June 1st (recommended period for the *Negotiated Water Allocation*), for the different years (including, in red, the years of the 2012-2020 drought period), where it is possible to obtain the flow to be released. ANA, within its sphere of competence, has placed itself in articulation with other bodies of the Federal Government, state governments and water resources management institutions, supply companies, city halls and basin committees in search of solutions that can minimize the effects of this severe drought.

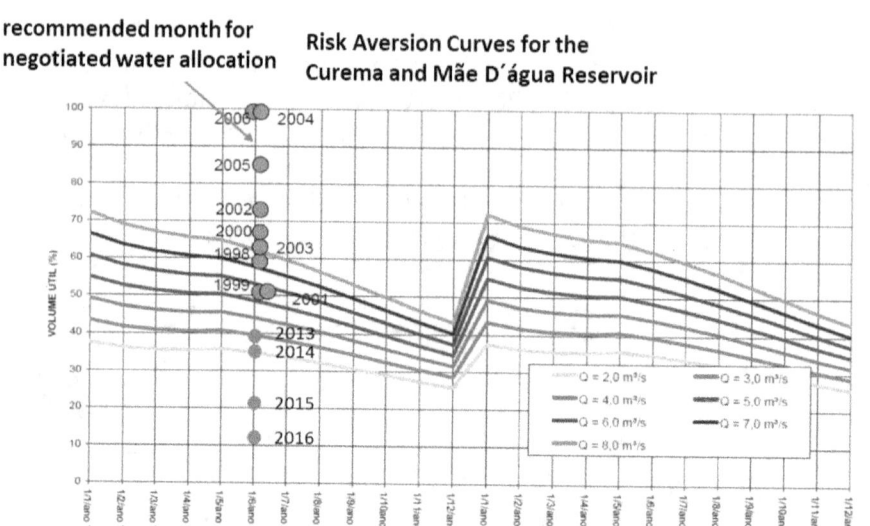

Figure 11: Risk Aversion Curves for the Curema and Mãe D´água Reservoir

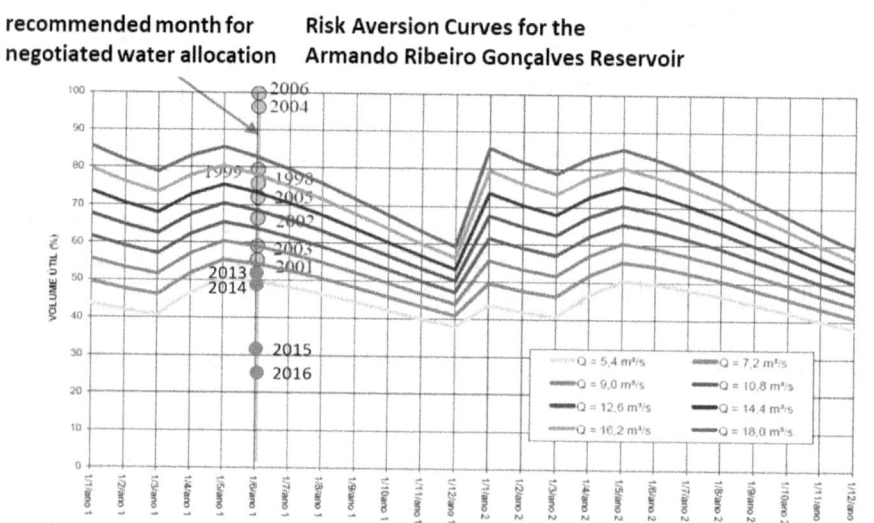

Figure 12: Risk Aversion Curves for the Armando Ribeiro Gonçalves Reservoir

4. Conclusions

With the increase in demand for water resources, there is a tendency to intensify conflicts with other users, especially during periods of long-term and severe droughts, such as that of 2012-2020. In situations like this, the integrated management of water resources stands out as an alternative to the resolution of conflicts, contributing to minimize the impacts of long periods of drought. In this article, the application of some of the techniques and methods developed, especially for semi-arid regions, in the hydrographic basin of the Piancó-Piranhas-Açu rivers were presented.

5. Bibliographic References

ANA/GEF/PNUMA/OEA (2004): Subprojeto 4.5C– Plano Decenal de Recursos Hídricos da Bacia Hidrográfica do Rio São Francisco-PBHSF (2004-2013) Estudo Técnico de Apoio ao PBHSF – Nº 16 ALOCAÇÃO DE ÁGUA. Alan Vaz Lopes; André Raymundo Pante; Leonardo Mitre Alvim de Castro; Marcos Airton de Sousa Freitas, 2004.

ANA (2016) - Agência Nacional de Águas (Brasil). Boletim de Acompanhamento dos Reservatórios do Nordeste do Brasil / Agência Nacional de Águas, Superintendência de Operações e Eventos Críticos. Brasília.

Freitas, M. A. S. (1995): Stochastische Abflussgenerierung in intermittierenden semiariden Gebieten (NO - Brasilien). Abschlussarbeit Weiterbildendes Studium Bauingenieurwesen, University of Hannover, Germany.

Freitas, M. A. S. (1996): Previsão de Secas por Meio de Métodos Estatísticos e Redes Neurais e Análise de Suas Características Através de Diversos Índices. In: IX Congresso Brasileiro de Meteorologia, 1996, Campos do Jordão. Anais do IX Congresso Brasileiro de Meteorologia, 1996.

Freitas, M. A. S. (2003): Alocação Negociada da Águas na Bacia Hidrográfica do Rio Gorutuba (Reservatório Bico da Pedra) - Minas Gerais. In: XV Simpósio Brasileiro de Recursos Hídricos, 2003, Curitiba. Anais do XV Simpósio Brasileiro de Recursos Hídricos. Porto Alegre: ABRH, 2003.

Freitas, M. A. S. (2009). A Regulação dos Recursos Hídricos. 1. ed. Rio de Janeiro: CBJE, 2009. v. 1. 174p.

Freitas, M. A. S. (2010). Que Venha a Seca: modelos para gestão de recursos hídricos em regiões semiáridas, Ed. CBJE, 1ª ed., Rio de Janeiro, 473p.

LIMA, C. A. G., CURI, W. F. & CURI, R. C. (2000). "Marco Regulatório para a Gestão do Sistema Curema-Açu e as Disponibilidades Hídricas do Reservatório Curema-Mãe D'água", RBRH — Revista Brasileira de Recursos Hídricos Volume 12 n.4 Out/Dez 2007, pp.73-86.

Freitas, M.A.S.; Billib, M.H.A. (1997): Drought prediction and characteristic analysis in semiarid Ceará, northeast Brazil. Sustainability of water resources under increasing uncertainty. Proceedings of an international symposium of the Fifth Scientific Assembly of the International Association of Hydrological Sciences (IAHS), Rabat, Morocco, 23 April to 3 May 1997, Rosbjerg, D. Boutayeb,

N.E.Gustard, A.Kundzewicz, Z.W.Rasmussen, P.F. (eds.).- Wallingford (United Kingdom): IAHS Press, 1997.- ISBN 0-901502-05-8.

Freitas, M. A. S.; Freitas, G. B. (2019a): The GAR(1) model with fragment method for hydrological drought risk assessment in semiarid regions. *Brazilian Journal of Development*, v. 5, p. 18267-18281, 2019.

Freitas, M. A. S.; Freitas, G. B. (2019b): On the Applicability of Multiseasonal Streamflow Generation Models for Intermittent Rivers. *IOSR JOURNAL OF MECHANICAL AND CIVIL ENGINEERING*, v. 16, p. 36-44, 2019.

Freitas, M. A. S.; Freitas, G. B. (2019c): Hydrological Drought Assessment: The Use of the ARRF Model for Monthly Streamflow Generation on Intermittent Rivers of the Northeast Brazil. *Quest Journals - Journal of Research in Environmental and Earth Sciences,* Volume 5, Issue 2 (2019) pp: 29-37 ISSN(Online):2348-2532.

Freitas, M. A. S. & Freitas, G. B. (2020): *Inteligência Artificial e Machine Learning: Teoria e Aplicações*, Março de 2020 (1ª edição), Independently published – Amazon, ISBN: 979-8621417178.

Freitas, M. A. S.; Gondim Filho, J. G. C. (2007). *"Curvas de Aversão ao Risco para os Reservatórios Armando Ribeiro Gonçalves e Curemas-Mãe D´água"* in Anais do XVII Simpósio Brasileiro de Recursos Hídricos, 2007, São Paulo.

Freitas, M. A. S., Lopes, A. V. (2003). *"A Avaliação da Demanda de Água para Irrigação: Aplicação à Bacia do Rio São Francisco"* in Anais do XIII CONIRD, Juazeiro, 2003.

Freitas, M. A. S.; Silveira, P. B. M.; Freitas, G. B. (2019): A Resilient Drought Risk Management Approach in the Semiarid Northeast Brazil. *INTERNATIONAL JOURNAL OF CURRENT RESEARCH*, v. 11, p. 6968-6974, 2019.

MARTINS, E. S. P. R.; BRAGA, C. F. C.; NYS, E.; SOUZA FILHO, F. A.; FREITAS, M. A. S. (2013). *Impacto das Mudanças do Clima e Projeções de Demanda Sobre o Processo de Alocação de Água em Duas Bacias do Nordeste Semiárido*. Banco Mundial, 1 ed., Brasília, Série Água Brasil 8, 112p.

MI (2000). *Relatório de Operação Integrada dos Açudes* - TOMO I.; IR. V/G. RT. GH. 003. Ministério da Integração Nacional, Secretaria de Infra-Estrutura Hídrica.

PLANO DE RECURSOS HÍDRICOS *Piranhas-Açu-Diagnóstico*, 2014, In: http://piranhasacu.ana.gov.br/produtos/sinteseDiagnostico.pdf

Resolução ANA nº 399/2004.

Resolução ANA nº 687/2004.

Rusteberg, B.; Freitas, M. A. S. (2018): IWRM Implementation in North-East Brazil (Results from WP 8). In: Abels, A., Freitas, M. A. S., Pinnekamp, J., Rusteberg, B. (Org.). Bramar Project - Water Scarcity Mitigation in Northeast Brazil. 1ed., Aachen: Department of Environmental Engineering (ISA), 2018, v. 1, p. 120-144.

CHAPTER 11

Definitions and Concepts relating to LMEO and the Delimitation of Permanent Areas of Protection with Water Function in the Light of the New Brazilian Forest Code

Marcos Airton de Sousa Freitas
Specialist in Water Resources at the National Water Agency - ANA; Ministry of Regional Development - MDR; masfreitas@ana.gov.br

Sandra Regina Afonso
PhD in Forestry Sciences - UnB; Researcher at the Brazilian Forest Service – SFB

Márcio Antônio Sousa da Rocha Freitas
Lawyer, Agronomist and Dr. Sc. (UFC); Prof. Univ. - Piauí State University (UESPI), Rua João Cabral, 2231 - Pirajá, Teresina - PI, 64002-150

Abstract: This article aims to present and discuss concepts related to the Annual Average Flood Line (LMEO) and the delimitation of the Permanent Preservation Areas (APP) that aim to protect water courses, considering the institution of the New Forestry Act, the 12.651 Law from May 25, 2012, also called the New Brazilian Forestry Code. The various concepts presented in a series of legal provisions relating to the delimitation of PPAs, aimed at protecting water courses, do not have enough clarity to enable the implementation of public policies in order to ensure the effective conservation of water resources. This is due to the various aspects discussed, as follows: i) the river is a dynamic system for which it is difficult to determine its regular river channel; ii) the wide range of ecosystems within the country for which the individual characteristics should be considered; and iii) the vast territorial extension that demands the use of remote analysis tools, which, in some cases, need to use simplifications in order to ensure its applicability.

Keywords: LMEO; Permanent Protection Area; Forest Code.

1. Introduction

This article aims to present and discuss concepts related to the demarcation of the Ordinary Flood Medium Line (LMEO in Portuguese) and the delimitation of Permanent Preservation Areas (APPs in Portuguese) that aim to protect water courses, considering the institution of the New Forest Law, the Law no. 12651, of May 25, 2012, also called the New Brazilian Forest Code.

According to Silva (2008) apud Oliveira & Miguez (2011), a river can be defined as "a dynamic system formed by the combination of two phases: a liquid phase represented by a basic flow with free, turbulent surface and deformable walls, governed by the laws of Fluid Hydraulics and Mechanics, and whose behavior determines the shape and geometry of the river channel; a solid phase, represented by a flow of solid particles of various dimensions and different physical-chemical and mechanical properties, generically called sediments, whose behavior can, in turn, modify the properties of the liquid stream". In this way, it is that through a feedback process, the flow changes the geometry of the river channel and the new configuration of this channel causes a change in some flow characteristics.

Thus, according to Oliveira & Miguez (2011), in conditions of natural balance, the "functioning of this dynamic system is responsible for the geometry and morphology of the rivers, determining their depths, widths, slopes, sinuosity of the watercourse and types of river configurations. bed. These properties show continuous fluctuations in time, whose average values over a sufficiently long period, are constant or vary on a small scale. In this case, it is said that the river is "**in regime**". Imbalance problems arise when modifying the watershed and / or introducing engineering works. Depending on these authors, the characterization of an **ordinary flood** can only be carried out when the river is in regime. The construction of reservoirs and the urbanization of the basin therefore change the conditions of the liquid phase and the solid phase and a new regime is produced, different from the previous one.

According to the Dictionary of Technical Terms for Irrigation and Drainage (ABID, 1978), of the Brazilian Association of Irrigation and Drainage, the term "**flood**", "**floods**" or "**flooding**" is defined as the "relatively high flow or level in a significantly higher than normal; also the flooding of lowlands that can result from it; body of water that rises, swells and floods land that is not normally covered by water ".

According to the floodplains, according to Christofoletti (1980), the river beds correspond to the spaces, which can be occupied by the flow of water: i) Tidal bed: which is included in the smaller bed and is used for runoff from low water; ii) **Smaller bed**: it is well delimited, fitted between generally well-defined margins; the flow of water in this bed is sufficiently frequent to prevent the growth of vegetation; iii) **Periodic or seasonal largest bed**: it is regularly occupied by floods, at least once each year; iv) **Exceptional largest bed**: where the highest floods, the floods flow; it is submerged at irregular intervals, but, by definition, not every year.

Imperial Law No. 1,507, of September 26, 1807, established the first notion of marginal lands in Administrative Law (Gasparine, 2006). The objective is to protect the marginal terrain of the navigable rivers of seven fathoms (15.40m). However, for the law to be applied, it was necessary to demarcate the midpoint of **ordinary floods**.

Law No. 1,507, of September 26, 1867, establishes, in its Art. 39, that "It is reserved for public easement on the banks of the navigable rivers and that the navigable ones are made, out of the reach of the tides, except for legitimate concessions made up to the date of publication of this law, the area of seven fathoms counted from the midpoint of ordinary floods to the interior, and the Government authorized to grant them in reasonable lots in the form of the provisions on the **navy lands**".

Decree No. 4,105, of February 22, 1868, advances in the definition of marginal and increased lands, according to Art. 1, § 2 "They are lands reserved for public easement on the banks of navigable rivers and of which are made navigable, all that bathed in the waters of the said rivers, out of the reach of the tides, go up to the distance of 7 **fierce fathoms** (15.4 meters) for the land part, counted from the midpoint of ordinary floods ".

Decree No. 24,643, of July 10, 1934, known as the Water Code, establishes, in its Article 11, § 2, that they are public Sunday, if they are not intended for common use, or by any legitimate title not belong to the private domain, the "land reserved on the banks of public currents in common use, as well as channels, lakes and ponds of the same species. Except for currents that are neither navigable nor floatable, they only compete to form others that are simply floatable, and not navigable". And in Art. 14, that "the reserved lands are those that, bathed by navigable currents, out of the range of the tides, go up to a distance of 15 meters to the part of the land, counted from the midpoint of ordinary floods".

Thus, Decree-Law No. 9,760, of September 5, 1946, states that:

> "Art. 1º Among the real estate of the Union are included:
> a)..
> b) the marginal lands of navigable rivers, in Federal Territories, if, by any legitimate title, they do not belong to private individuals;
> c) the marginal lands of rivers and the islands located in them, in the border strip of the national territory and in the areas where the influence of the tides is felt; "

"Art. 4º Marginal lands are those that bathed by navigable currents, out of the range of the tides, go up to a distance of 15 (fifteen) meters, measured horizontally for the part of the land, counted from the average line of ordinary floods. "
"Art. 9º It is the competence of the Union Patrimony Service (SPU) to determine the position of the average level water lines for the year 1831 and the average of ordinary floods."

Furthermore, the Constitution of the Federative Republic of Brazil, of October 5, 1988, establishes, in its Art. 20, that, among others, are assets of the Union "the lakes, rivers and any water currents in lands of its domain, or that they bathe more than one State, serve as limits with other countries, or extend to foreign territory or come from it, as well as marginal lands and river beaches ".

Silva (2008) clarifies that "the 1988 Constitution, which in its article 20, III, in an unprecedented manner, expressly mentioned marginal lands immediately after the enunciation of the waters owned by the Union, including in such ownership the currents of water in lands of its domain, or that bathe more than one State, serve as limits with other countries, or extend to foreign territory or come from it, so that only the lands located in the margins of these water currents would belong to the Union". Thus, according to Silva (2008), the device makes it clear that federal rivers would be those framed in the criteria of the device, whether navigable or not, as it speaks of "water currents", in such a way that even the lands located on the banks of currents do not waterways appear to be included among the Union's assets.

Regarding the Permanent Preservation Areas - APP, Federal Law No. 4,771 / 1965 (Forest Code) stated:

"Art. 2 ° For the sole purpose of this Law, forests and other forms of natural

vegetation are considered to be permanently preserved:

a) along rivers or any watercourse from its highest level in a marginal band whose

minimum width will be: (text by Law No. 7,803 of 7/18/1989)

1 - 30 (thirty) meters for water courses less than 10 (ten) meters wide; (Wording

given by Law No. 7,803 of 7/18/1989)

2 - 50 (fifty) meters for water courses that are 10 (ten) to 50 (fifty) meters wide;

(Wording given by Law No. 7,803 of 7/18/1989)

3 - from 100 (one hundred) meters for water courses that are 50 (fifty) to 200 (two

hundred) meters wide; (text by Law No. 7,803 of 7/18/1989)

4 - from 200 (two hundred) meters for water courses that are 200 (two hundred)

to 600 (six hundred) meters wide; (text by Law No. 7,803 of 7/18/1989)

5 - 500 (five hundred) meters for water courses that are wider than 600 (six

hundred) meters; (Included by Law No. 7,803 of 7/18/1989)"

Xavier et al. (2011), citing work by Miranda et al. (2008), state that there are three major difficulties in mapping and quantifying the areas of permanent preservation linked to hydrography. The first would be the lack of homogeneous and detailed mapping of the hydrographic network in Brazil, especially in the Amazon. The second, still according to these authors, concerns Resolution 303/2002, of the National Environment Council - CONAMA, which establishes as an area occupied by the river, to calculate its width, not the permanently flooded bed, but the strip floodable "from the highest level", which is defined by the same resolution as the "level reached during the seasonal flood of the perennial or intermittent watercourse". Finally, the third difficulty lies in adjusting the marginal strips, on a case-by-case basis, since this floodable area must be added with a marginal band varying from 30 to 500 m on each side, depending on the width of the maximum flood area.

According to CONAMA Resolution No. 303, of March 20, 2002, which provides for parameters, definitions and limits of **Permanent Preservation Areas - APP**, we have:

Art. 2 For the purposes of this Resolution, the following definitions are adopted:

I - highest level: level reached during the seasonal flood of the perennial or

intermittent watercourse;

Art. 3 The Permanent Area constitutes the area located:

I - in marginal band, measured from the highest level, in horizontal projection,

with minimum width, of:

a) thirty meters, for the watercourse less than ten meters wide;

b) fifty meters, for the watercourse ten to fifty meters wide;

c) one hundred meters, for the water course fifty to two hundred meters wide;

d) two hundred meters, for the watercourse two hundred to six hundred meters

wide;

e) five hundred meters, for the watercourse more than six hundred meters wide.

CONAMA Resolution No. 341, of September 25, 2003, published in DOU No. 213, of November 3, 2003, Section 1, page 62, resolves in its Article 1 to add to CONAMA Resolution No. 303, of March 20, 2003 2002, published in the Official Gazette, of May 13, 2002, Section 1, page 68, the following considerations: "Considering the convenience of regulating arts. 2 and 3 of Law No. 4,771, of September 15, 1965, with regard to the Permanent Preservation Areas; Considering that it is the duty of the Public Power and individuals to preserve biodiversity, notably flora, fauna, water resources, natural beauty and ecological balance, avoiding water, soil and air pollution, an intrinsic assumption for the recognition and exercise of the right property, pursuant to arts. 5, caput (right to life) and item XXIII (social function of property), 170, VI, 186, II, and 225, all of the Federal Constitution, as well as of art. 1.299, of the Civil Code, which obliges the owner and squatter to respect the administrative regulations".

The Normative Guidance, which governs the demarcation of marginal lands and their additions - ON-GEADE-003, of June 4, 2001, indicates that for the calculation of the average of ordinary floods, "the maximum annual quotas referring to to floods with a recurrence period of 3 years, excluding floods with recurrence periods of 20 years or more ". Additionally, that "the use, for calculating the average of ordinary floods, of quotas referring to floods with a recurrence period of more than 3 years will be allowed, as long as duly justified". And that "only data from fluviometric stations that have, at least, 20 years of observations will be used". It goes on to affirm that "in possession of the form containing the observation data of the floods of a given fluviometric station, the maximum annual quotas should be listed in decreasing order. The quotient obtained by dividing the number of years of observation in a fluviometric station by the recurrence period (in years), will indicate the number of flood shares with recurrence periods equal to or greater than that used as a reference for the calculation ". Continuing, he states that floods with a recurrence period of less than 3 years and equal to or greater than 20 years will be discarded. And, finally, that "the average of the ordinary floods of a fluviometric station will be the arithmetic average of the maximum annual quotas for the floods with recurrence periods between 3 and 20 years".

However, according to Oliveira & Miguez (2011), "the use of floods with a recurrence period longer than two years is incompatible with the spirit of Law No. 1,507, of September 26, 1807, contradicts the etymological meaning of the word ordinary and does not find support in usual procedures in the study of water resources and in the definitions of the river bed in the international technical literature ". And they go on to affirm that "the demarcation could still be supported by modern hydrology procedures, with the use of mathematical models, which point out, through calculation, the probable current bankfull line and the associated flow. Law No. 1,507, of 1807, needs adjustments, as it is not reasonable to indicate the probable bankfull line in 1807, as it was not possible to faithfully reproduce what happened at that time, due to lack of records. Furthermore, and more importantly, it is also unreasonable to use a reality that no longer exists by reference".

According to the Manual of Land Regularization in Lands of the Union, of the Secretariat of Patrimony of the Union - SPU (Saule Júnior et al., 2006), floodplains "are areas located along rivers with annual cycles, marked by periods of floods and ebb. They are lands that, periodically, are flooded during a flood of the river and discovered with the ebb". Thus, according to the aforementioned manual, due to the fact that there is no concept of floodplain described by law, the concept of a larger bed, adopted by CONAMA Resolution No. 004/1985, is adopted as a basis, that is, "enlarged or larger gutter of a river, occupied during annual periods of flood ".

Morais et al. (2011) presented a bibliographic review on the functions considered most relevant from the hydrological point of view and the protection of the water body performed by riparian zones that, according to these authors, are generally contained within APPs, in view of the current impacts on water bodies, especially with regard to the erosive processes of the river channel, the reduction of runoff and the supply of sediments and nutrients in the water body. Additionally, the hydrological function of the riparian zone is divided into several factors that directly influence the stability of the hydrographic basin and indirectly on the quality of the water, among which they mention: attenuation of the flood peak; dissipation of energy from runoff due to vegetation roughness; thermal water balance; margin stability; nutrient cycling; and sedimentation control.

Victoria & Mello (2011) commenting on several studies in order to identify the Permanent Protection Areas - APP, assess that, these estimates come up against limitations, mainly related to the lack of adequate cartographic information. And that, in the case of APPs on the banks of rivers, these dimensions depend on the width of the water courses in the highest flood level and that, "APP estimates for large basins that disregard the width of the drainage channel are subject to significant uncertainties".

Thus, if the different legal provisions are not clear in the sense of defining unequivocally, both the Average Line of Ordinary Floods - LMEO, and the Permanent Protection Areas - APP, there is no doubt about the legal competence of each one of them. The body responsible for promoting the necessary actions to identify, demarcate, discriminate, register, register and inspect real estate owned by the Federal Government is the Federal Heritage Secretariat - SPU, of the Ministry of Planning, Budget and Management. With Federal Law No. 5,972 / 1973, the Federal Government was obliged to register its assets. And, from 1998 onwards, based on federal legislation on the administration of the Union's real estate, it was mandatory to register these assets in the Real Estate Registry Office.

Law No. 12,651, of May 25, 2012, provides for the protection of native vegetation; amends Laws 6,938, of August 31, 1981, 9,393, of December 19, 1996, and 11,428, of December 22, 2006; repeals Laws 4,771, of September 15, 1965, and 7,754, of April 14, 1989, and Provisional Measure 2,166-67, of August 24, 2001; and makes other arrangements.

Depending on Art. 3, item II, of the aforementioned Law, the definition of Permanent Preservation Area - APP: protected area, covered or not by native vegetation, with the environmental function of preserving water resources, the landscape, geological stability and biodiversity, facilitate the gene flow of fauna and flora, protect the soil and ensure the well-being of human populations.

According to Art. 4, it is considered a Permanent Preservation Area, in rural or urban areas, for the purposes of this Law:

I - the marginal strips of any natural watercourse, from the edge of the regular bed gutter, in a minimum width of:

a) 30 (thirty) meters, for water courses less than 10 (ten) meters wide;

b) 50 (fifty) meters, for water courses that are 10 (ten) to 50 (fifty) meters wide;

c) 100 (one hundred) meters, for water courses that are 50 (fifty) to 200 (two hundred) meters wide;

d) 200 (two hundred) meters, for water courses that are 200 (two hundred) to 600 (six hundred) meters wide;

e) 500 (five hundred) meters, for water courses that are wider than 600 (six hundred) meters.

Item XIX, of Art. 3, defines a **regular bed**, as "the trough through which the waters of the watercourse regularly flow during the year".

Thus, there was a change in the determination of the marginal range, which is now measured from the edge of the regular gutter, instead of being measured from the highest level, according to Art. 2 of Federal Law No. 4,771 / 1965, revoked. Depending on CONAMA Resolution No. 303, of March 20, 2002, the highest level would be the level reached during the seasonal flooding of the perennial or intermittent watercourse, in horizontal projection.

According to several researchers from the Brazil Committee in Defense of Forests and Sustainable Development (2012), as well as Andrade (2015), this change can compromise the environmental balance, both in humid areas and in semiarid regions.

The rivers of the Brazilian semiarid region present high flow variability, influenced by the rain dynamics, which are generally concentrated and poorly distributed, both temporally and spatially. In the dry season, rivers dry up from their headwaters until close to the coast, which does not happen with rivers in more humid regions (Ab'Saber, 2003; Freitas, 2010; Andrade, 2015).

According to Andrade (2015), for semi-arid rivers this change significantly reduces their APPs, as the regular bed has a considerably smaller size in relation to the larger bed, in addition to some of the rivers drying out during the dry season, which makes it difficult to define the APP . Another aggravating factor is that most of the semi-arid rivers have their flow controlled by dams, which camouflages the regular bed, making it even smaller. Thus, changes in the form of delimiting the width of the APP of the rivers, especially the semi-arid rivers may have their protection areas compromised, contributing to changes in their river dynamics, with the increase of the erosion processes of the banks and the silting of the fluvial channel.

For humid areas, such as the Pantanal, Piedade et al. (2012) evaluate that the reference to the width of the regular gutter does not address the most important aspect in these systems, which is the extension and lateral expansion of these wet areas, which varies over the landscape and the year. According to the authors, APPs should be delimited from the highest level of the flood in the humid areas of the national territory.

Law 12.651 / 2012 also establishes, in its article 29 that all rural properties must be registered in the Rural Environmental Registry, in which the location, among others, of the Permanent Preservation Areas must be informed:

Art. 29. The Rural Environmental Registry - CAR is created, within the scope of the

National Environmental Information System - SINIMA, a national electronic public register, mandatory for all rural properties, with the purpose of integrating the environmental information of the properties and rural possessions, composing a database for control, monitoring, environmental and economic planning and combating deforestation.

Paragraph 1. The registration of the rural property in the CAR should preferably be done at the municipal or state environmental agency, which, under the terms of the regulation, will require the owner or rural owner:

I - identification of the rural owner or owner;

II - proof of ownership or possession;

III - identification of the property by means of a plan and descriptive memorial, containing the indication of the geographical coordinates with at least one mooring point on the perimeter of the property, informing the location of the remnants of native vegetation, the Permanent Preservation Areas, the Use Areas Restricted, the consolidated areas and, if any, also the location of the Legal Reserve.

The Rural Environmental Registry has been representing an important and innovative tool, both to subsidize public policies and for producer planning. In a very simple way, when registering your property, the owner declares only the width of the river and from that the system generates the delimitation of the water APPs of your property, without entering into the merit regarding its measurement method.

2. Final Considerations

The various concepts presented in a series of legal provisions related to the delimitation of APPs that aim to protect water courses are not clear enough to allow the implementation of public policies capable of guaranteeing the effective preservation of water resources. This is due to the various aspects discussed, namely: i) the river represents a dynamic system for which it is difficult to determine the regular bed channel; ii) the wide range of ecosystems in the country for which their specificities must be considered; and iii) the vast territorial extension that requires the use of remote analysis tools that, in some cases, need simplifications to guarantee their applicability.

3. Bibliographic References

AB'SABER, A.N. (2003). *Os domínios de natureza no Brasil – potencialidades paisagisticas*. São Paulo: Atiliê Editorial, 159p.

ANDRADE, J. H. R. (2015). *"Mudanças na Forma de Delimitar a Área de Proteção Permanente (APP) dos Rios e Suas implicações para os Rios Semiáridos"*. In II Workshop Internacional sobre Água no Semiárido Brasileiro, Campina Grande, PB.

CHRISTOFOLETTI, A. (1980). *Geomorfologia*. São Paulo. Edgard Blücher, 2ª edição, 83 p.

FREITAS, M.A.S. (2010). *Que Venha a Seca*, Ed. CBJE, Rio de Janeiro, RJ, 473p.

GASPARINE, D. (2006). *Direito Administrativo*. 11ªed.rev.atualizada. São Paulo: Saraiva, 876 p.

MIRANDA, E. E., OSHIRO, O. T., VICTORIA, D. C., TORRESAN, F. E. & CARVALHO, C. A. (2008). **O Alcance da Legislação Ambiental e Territorial**. AgroAnalysis, FGV. Disponível em: http://www.agroanalysis.com.br/especiais_detalhe.php?idEspecial=35&ordem=2

MORAIS, A., GONÇALVES, L. P., ROSA, E. U. & COSTA, S. R. A. (2011). *"Eficiência da vegetação Ripária na Faixa Marginal de Proteção (APP de Margem de Rio)"* in Anais do XIX Simpósio Brasileiro de Recursos Hídricos, Maceió, Alagoas, 2011.

OLIVEIRA, R. C. N. de & MIGUEZ, M. G. (2011). *"O Domínio dos Terrenos Marginais e seu Impacto na Requalificação Fluvial"* in Anais do XIX Simpósio Brasileiro de Recursos Hídricos, Maceió, Alagoas, 2011.

PIEDADE, M.T. F.; JUNK, W. F.; SOUSA JR, C. C. C.; SCHÖNGART, J.; WITTMANN, F.; CANDOTTI, E.; GIRARD, P. (2012). *As áreas úmidas no âmbito do Código Florestal Brasileiro*. In: COMITÊ BRASIL EM DEFESA DAS FLORESTAS E DO DESENVOLVIMENTO SUSTENTÁVEL. Código Florestal e a Ciência: o que nossos legisladores ainda precisam saber. Comitê Brasil. Brasília-DF.

SAULE Júnior, N. e outros. *Manual de Regularização Fundiária em Terras da União*. Organização de Nelson Saule Júnior e Mariana Levy Piza Fontes. São Paulo: Instituto Polis; Brasília: Ministério do Planejamento, Orçamento e Gestão, 2006. 120p.

SILVA, M. L. (2008). *"Dos Terrenos Marginais da União: conceituação a partir da constituição federal de 1988"*. Revista da AGU – Eletrônica nº 82, nov.2008.

VICTORIA, D. C. & MELLO, J. S. (2011). *"Avaliação de Diferentes Métodos para Estimativa de Áreas Marginais de Cursos D´água na Bacia do rio Ji-Paraná (RO)"* in Anais do XV Simpósio Brasileiro de Sensoriamento Remoto – SBSR, Curitiba, PR, Brasil, 2011, INPE, p.3890.

XAVIER, M. C., CARVALHO Jr., M., VIEIRA, M., MENEZES, G. & MOREIRA, G. (2011). *"Metodologia para Demarcação de Faixa Marginal de Proteção de Curso D'água (APP de Margem de Rio) no Estado do Rio de Janeiro"* in Anais do XIX Simpósio Brasileiro de Recursos Hídricos, Maceió, Alagoas, 2011.

CHAPTER 12

Rainfall-Runoff Model CN3S for Hydrological Drought Risk Management in Brazilian River Basins

Marcos Airton de Sousa Freitas
Specialist in Water Resources at the National Water Agency - ANA; Ministry of Regional Development - MDR; masfreitas@ana.gov.br

Abstract: The CN-3S (Curve Number 3 Step) model, with a wide history of application in the semi-arid region of Northeast Brazil, is based on the relationships developed by the US Conservation Service of the CN (Curve Number) curves and it is composed of six calibration parameters. It is a conceptual model for streamflow generation and it has been applied to several hydrological studies in the semi-arid region of Piauí, Ceará, Pernambuco and Paraíba States. The CN-3S deterministic rainfall-runoff model (Curve Number with Three Step Antecedent Precipitation) has been developed with the objective of generating synthetic streamflow for the simulation of reservoir operations and could be use for hydrological drought risk management.

Keywords: CN-3S model; Rainfall-Runoff Model; Hydrological Drought semi-arid region.

1. Introduction

Since the first quantitative studies in Hydrology, researchers have come across the rainfall-runoff models. During the first half of the 19th century, records of flow measurements from several European rivers were published, an initial milestone for the development of hydrological process models (Linsley, 1981). The first models had their origins, mainly, in response to urban sewage problems, drainage systems design and dam spillway design (Todini, 1988).

From a historical perspective, it is important to emphasize the use of the rational method from the beginning of the last century. Then came the concept of unitary hydrograph (Sherman, 1932), based on the principle of superposition, until, in the 50's of the last century, the first so-called "conceptual" models have been appeared.

The United States Soil Conservation Service, in 1954, has presented a procedure to estimate direct runoff from precipitation based on so-called Curves Number or CN. Several models have been emerged, that have been used such curves. After the 1960s, due to the incorporation of the computational tool, rainfall-runoff models gained a great boost, until they reached the current state of the art.

Rainfall-Runoff Models

A model is called rainfall-runoff (MCV) when, starting from certain data, in general, precipitation and evaporation, it is possible to obtain, through empirical and / or physical equations, the streamflow in a determined section of a hydrographic basin. In Viessman et al. (1977), as well as in Fleming (1975) there exist descriptions of several hydrological models.

As models are simplified representations of the real system (nature), the model that incorporates greater complexity is not always the most suitable for a given use, since the availability of data must be considered. Therefore, it is essential to bear in mind the purpose of using the model. Thus, for a perfect choice of the model to be used, the hydrologist should observe, apart, of course, the availability of the model, two other aspects: a) the purpose of using the model and b) the reliability of the results as a whole: data, model, calibration, validation and application (Freitas & Porto, 1990).

Since the model is a simplistic representation of the complex reality, the appearance of imperfections is notorious. Therefore, there is an urgent need to pay attention to the various sources of uncertainty. O'Donnell & Canedo (1980) listed the main sources of uncertainty, namely: records of hydrometeorological data from the hydrographic basin, structures of the rainfall-runoff models and calibration of rainfall-runoff models.

Retnam & Willian (1988) report studies on errors resulting from the use of point measurements of an input variable of continuous into space. Wood et al. (1988) study the effects of spatial variability and scale problem in the case of hydrological models. Delhomme (1979), Seven et al. (1989) and Smith & Herbert (1979) analyze the spatial variability and uncertainties of model parameters. Problems related to the structure of the models derive mainly from the imperfect knowledge of the physics of hydrological processes. The processes of infiltration and percolation of water in the soil are those that, generally, prevent the obtaining of better results in the use of rainfall-runoff models.

Tucci et al. (1988), trying to solve this problem related to infiltration, propose algorithms that consider the spatial variability of the infiltration capacity and that can be incorporated into the rainfall-runoff models. Another interesting aspect raised by Sorooshian & Gupta (1985) deals with the structural identification of models. These authors focus on a procedure to determine if the model structure is an aid or an obstacle for determining the optimal values of the parameters. Gupta & Sorooshian (1985) discuss the threshold parameters, which lead the model to operate in different neighborhood modes, depending on its state. Canedo et al. (1989) alleviate this problem with the use of smoothing techniques.

Pilgrim et al. (1982) have presented an important study on the effects of the size of the hydrographic basin on the flow relationships. According to Sorooshian et al. (1983) the objective function to be employed must be such as to efficiently abstract the information contained in the data. Diskin & Simon (1977) proposed a systematic procedure for the selection of objective function, but it involves a lot of subjectivity. The optimization methods can be classified into direct and indirect methods. Indirect methods can also be subdivided into 1st order and 2nd order methods. Discussions about these methods can be found in Kuester & Mize (1974), Gupta & Sorooshian (1983), as well as in Bard (1974).

Models Classification

Clarke (1973) has allocated the mathematical models in Hydrology into four groups: i) stochastic-conceptual; ii) stochastic-empirical; iii) deterministic-conceptual; and iv) deterministic-empirical. However, each one of them can be subclassified in concentrate or distributed. A model is said to be concentrated when it ignores the spatial distribution of the input variables and parameters that characterize physical processes, varying only in relation to time. In the case of distributed models, this spatial distribution is taken into accounting.

A model is said to be stochastic if for a given input value there is a probability or element of randomness related to the input data. Conceptual and empirical models differ basically by the fact that those incorporate intrinsically into their procedural relationships physical, chemical, biological and other formulations, whereas in empirical or "black box" models this does not occur.

Another classification was given by Todini (1988), who allocated the models, according to their structure, in: i) purely stochastic; ii) concentrated integral; iii) distributed integral; and iv) distributed differential. Regarding the parameters, the models could be classified as stochastic or physical.

Application Fields of Rainfall-Runoff Models

In line with Kuczera (1983), the application fields of rainfall-runoff models are: i) extension of streamflow series; ii) generation of streamflow statistics; iii) access to the effects of changes in land use; iv) streamflow prediction in ungauged basins and v) prediction of the effects arising from land use changes in hydrological regimes.

According to Canedo (1989), the potential applications of rainfall-runoff models are: i) in the planning and management of watersheds; ii) in the hydrological projects of construction and reservoirs operation; iii) in irrigation or drainage projects; and iv) in studies of floods and droughts. The steps for using a rainfall-runoff model can be summarized in four items, namely: i) choice or formulation of the model; ii) calibration; iii) validation; and iv) model application.

The models used here have been for a lot of time researched and used by various public and private organizations in reservoir design and water availability studies in hydrographic basins in the Brazilian semi-arid region.

CN-3S Model

The CN-3S (Curve Number 3 Step) model, with a wide history of application in the semi-arid region of Northeast Brazil, is based on the relationships developed by the US Conservation Service of the CN (Curve Number) curves and it is composed of six calibration parameters. It is a conceptual model for streamflow generation that has been developed by Taborga & Freitas (1987) and it has been applied to several hydrological studies in the semi-arid region of Piauí, Ceará, Pernambuco and Paraíba States. Table 1 shows some applications of the model to the Northeast basins, according to Freitas & Porto (1991). Figure 1 shows the CN-3S model version implemented on an Excel spreadsheet.

The CN-3S deterministic rainfall-runoff model (Curve Number with Three Step Antecedent Precipitation) has been developed with the objective of generating synthetic streamflow for the simulation of reservoir operations. The CN-3S uses, as input data, necessary to calculate the runoff depth for a given time interval, the rainfall of the period itself and the precipitation of the three previous periods.

For the evaluation of the value of this flow sheet, the equations of real flow with the potential flow were used, according to the equations of the U. S. Conservation Service:

$$\frac{(P - I_m) - Q}{S} = \frac{Q}{(P - I_m)}$$

(1)

where:
P - rainfall height (mm);
Q - direct flow (mm);
S - maximum potential difference between P and Q;
Im - interception, infiltration and superficial storage.

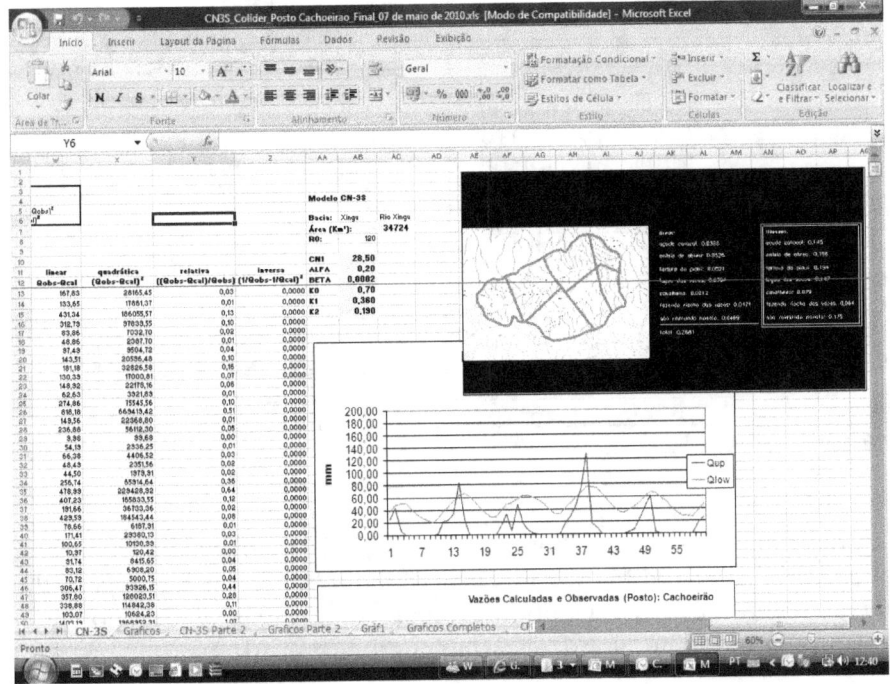

Figure 1. CN-3S model version implemented on an Excel spreadsheet

Taking the value of Q in the previous expression, with Im = ALFA * S, results in:

$$Q = \frac{(P - ALFA * S)^2}{P + (1 - ALFA) * S}$$

(2)

The U. S. Conservation Service technicians adopt ALFA equal to 0.2, that is, 20% of the soil's storage capacity. For the CN-3S model, however, ALFA is a calibration parameter. In the several analyzed cases (Freitas & Porto, 1990), the calibration, adopting the value of ALFA equal to 0.2, proved to be satisfactory, which reduces the number of parameters to be adjusted to five.

Table 1: Applications of the CN-3S model to the basins of Northeast Brazil

HYDROGRAPHIC BASIN	PEDRA REDONDA	GROAIRAS	MATRIZ	ANTENOR NAVARRO	PAJEÚ
area (Km²)	3340	2759	468	1257	6170
state	Piauí	Ceará	Pernambuco	Paraíba	Ceará
calibrated parameters					
CNI	24.0	26.5	21.1	15.5	16.5
ALFA	0.2	0.2	0.225	0.2	0.2
BETA	0.00300	0.00135	0.00260	0.00390	0.00600
K0	0.57	0.88	1.00	0.69	0.25
K1	0.0105	0.0110	0.0090	0.0180	0.0600
K2	0.42	0.95	0.95	0.40	0.40
Observed series					
mean (mm)	2.02	8.50	2.19	4.59	6.20
standard deviation (mm)	6.01	31.12	7.40	15.98	20.08
Assymetry coef.	4.13	6.09	5.11	4.34	5.69
calculated series					
mean (mm)	2.05	8.33	2.15	4.59	6.86
standard deviation (mm)	5.83	31.43	7.24	12.75	18.98
Assymetry coef.	3.98	5.92	5.29	3.85	6.54
monthly correlation coef.	0.916	0.978	0.983	0.921	0.956

According to details found in Taborga & Freitas (1987), from the numerical values of the CNs tabulated by the SCS, according to the characteristics of the previous rain and the soil-vegetation complexes, calculating the potential regression curves, we arrive at the following multiple regression:

$$CNV_j = 0.925 * CNI^{1.019} * V_j^{8.256 - 0.479 * \ln(CNI)}$$

(3)

where CNI is a model adjustment parameter.

The value of Vj in the above equation expresses the antecedent precipitation coefficient, using as input the rainfall of the three times intervals preceding the interval in question. The values of Vj ($1 \le$ Vj ≤ 3), are computed by the following expression:

$$V_j = 1 + BETA*(P_{j-1} + K0*P_{j-2} + K0^2 * P_{j-3})$$ (4)

where:

BETA e K0 – adjustment parameters of the antecedent precipitation,

With the CNVj value obtained, then the Sj value is calculated using the following equation:

$$CN = \frac{1000}{(S/25.4)+10}$$ (5)

The previous equation represents the relations of the CN curves (Curve Number), depending on the variable S, that is, the maximum potential difference. The slide for the direct flow Qup is obtained by Equation 1, with the precipitation values P and the parameter ALFA.

The CN values, depending on the antecedent precipitation and the land use and class. To find the relationship between the data (V = 1, V = 2 and V = 3), a cross regression can be used. A regression equation was adjusted for each column in Table 2. The columns of the values of x and y contain the coefficients and the exponents between V = 1, V = 2 and V = 3, for each value of CN, according to the equation described below:

$$CNV = x.V^y$$ (6)

Afterwards, a regression is performed between the pivot column (CN-I) with the coefficients shown in column x.

$$x = 0.925(CN1)^{1.019}$$ (7)

Between the pivot column (CN-I) and the y column, a logarithmic regression given by:

$$y = 2.356 - 0.478.\ln(CN1)$$ (8)

Table 2. CN values and regression coefficients

V=1 CN-I	V=2 CN-II	V=3 CN-III	x	y
4	10	26	3.75	1.663
7	15	33	6.65	1.378
9	20	39	8.73	1.315
12	25	45	11.72	1.188
15	30	50	14.76	1.086
19	35	55	18.73	0.958
23	40	60	22.72	0.865
27	45	65	26.72	0.793
31	50	70	30.73	0.736
35	55	75	34.76	0.689
40	60	79	39.77	0.616
45	65	83	44.80	0.554
51	70	87	50.75	0.483
57	75	91	56.72	0.423
63	80	94	62.80	0.362
70	85	97	69.80	0.295
78	90	98	77.99	0.208

Substituting both above equations in Equation 6, results in:

$$CNV_j = 0.925(CN1)^{1.019}.V_j^{2.356-0.478\ln(CN1)}$$

$$(9)$$

The regression coefficient of this equation is 0.997. The values of Vj in the equation correspond to the coefficient of the monthly antecedent precipitation (Cordery, 1970):

$$V_j = 1 + BETA(N_{j-1} + K0.N_{j-2} + K0^2.N_{j-3})$$

$$(10)$$

V_j = antecedent precipitation coeficiente ($1 \leq V_j \leq 3$),
BETA= antecedent precipitation parameter;
K0 = antecedent precipitation parameter.

Base flow or basic flow is calculated under the assumption that a portion of the difference between precipitation and the direct runoff, at a K1 rate, feeds the water table, and it in turn undergoes depletion at a rate K2, corresponding to the basic flow Qlow (mm), according to the sequence below:

$$R_j = R_{j-1} + K1*(P_j - Q_{up}), \quad K1<1$$

$$(11)$$

$$Q_{low} = K2*R_j$$

$$(12)$$

The water table after Qlow depletion, at the end of period j, results in:

$$R_j = R_{j-1} + K1*(P_j - Q_{up}) - Q_{low} ,$$ (13)

Then, the total streamflow is given by:

$$Q_{total} = Q_{up} + Q_{low}$$ (14)

As an initial condition of R0, the value of zero can be adopted for non-perennial rivers and for perennial rivers, it is calculated by iteration, until reaching the amount of the first observed flow value.

CN-3S Model Application

For the use of the CN-3S model, the calibration / validation process is initially required, which consists of an iterative process, where, starting from the available knowledge of the model parameters and, using the search algorithm, called the Rosenbrock Method (1960), embedded in the CN-3S model, the search for the set of parameters that best describes the behavior of the basin.

The collection and documentation of rainfall data is of strategic importance in the practice of water resource management. In the case of the region under study, data were obtained from SUDENE covering the period from 1963 to 1983 (Itaim River Basin, at the Maria Preta station). These data are contained in the Canindé / Piauí Rivers Hydrographic Basin Master Plan, having been homogenized by the DNAEE MSDHD PROHD program, based on the accumulated mass double curve.

Similar as pluviometry, the collection and treatment of streamflow data has, over the years, suffered from institutional changes that have caused damage to access to this data. The Maria Preta station (34450000), on the Itaim river, is operated by DNAEE, with latitude 07 ° 32 'S and longitude 41 ° 20' W, starting operation on May 1, 1967. The situation of the Maria Preta station, according to the Canindé / Piauí River Basin Water Resources Master Plan, is as follows: i) Period with constant daily quota data (with failures): 05/27/1967 to 05/31/1978 and April 1, 1998 to September 30, 1997; ii) period with average daily discharges consisting of DNAEE (with failures): 05/27/1967 to 05/31/1979 and 04/01/1998 to 12/31/1999.

The period chosen for the calibration was from 1967 to 1971, obeying one of the most important criteria when selecting a period for calibration: that includes the largest flood peaks and the largest recorded droughts. The obtained parameters are presented in Table 3, while in Figures 2 and 3 the results of the calibration and verification can be observed.

Table 3. Parameters of the CN-3S model calibrated for the Itaim river basin, at the Maria Preta station (34450000)

Period/ Parameters	CNI	ALFA	BETA	K0	K1	K2
1967-1976	23.700	0.200	0.003	0.940	0.010	0.330

CNI = coefficient related to the soil-vegetation complex

ALFA = parameter related to interception and infiltration

BETA = adjustment parameter related to the antecedent precipitation

K0 = adjustment parameter related to the antecedent precipitation

K1 = water table rate coefficient

K2 = groundwater depletion rate coefficient

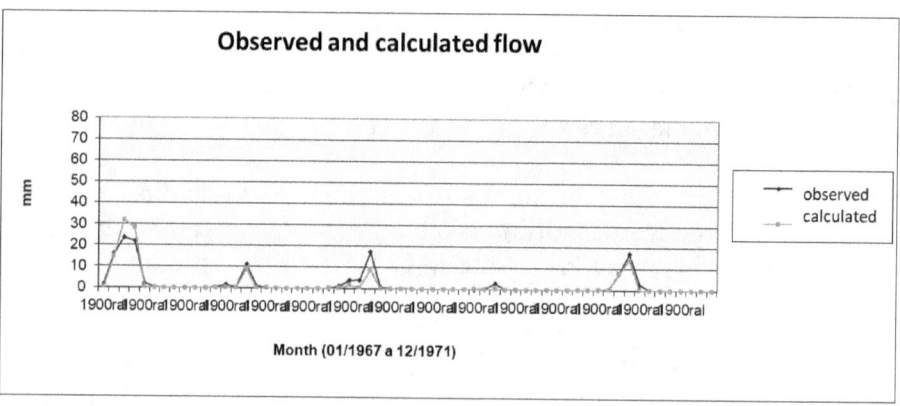

Figure 2. Observed and Calculated streamflow (Calibration Period) for the Maria Preta station

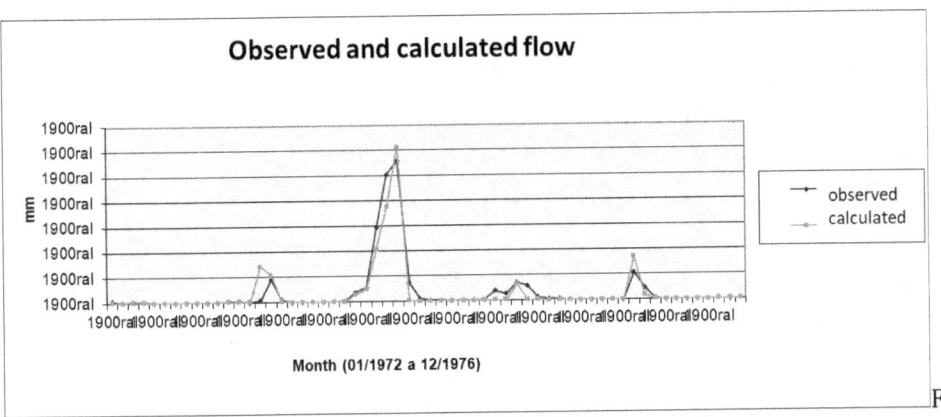

Figure 3. Observed and Calculated streamflow (Validation Period) for the Maria Preta station

For the calibration period, the adjustments were represented with good precision, particularly the low flows, showing that the parameters that reflect the emptying process of the different surface and underground reservoirs reached adequate values. It was also possible to reasonably reflect the maximum flood flows observed, with the complex peak of 1967 being represented satisfactorily and the peaks of the years 1968, 1969 and 1971, quite accurately.

The validation process extended to the 1972-1976 period, with very favorable results. The droughts were simulated in an acceptable way, some being slightly underestimated and others slightly overestimated. The main maximum flow rate, corresponding to 1974, is reproduced with extreme precision.

The model demonstrated that, despite its simple formulation, it is efficient in reproducing the monthly runoff in hydrographic basins in Brazil, thus serving as an instrument to support water management in the country. Figure 4 shows the observed and calculated streamflow curves (Calibration Period) and streamflow permanence curves for the Cachoeirão station.

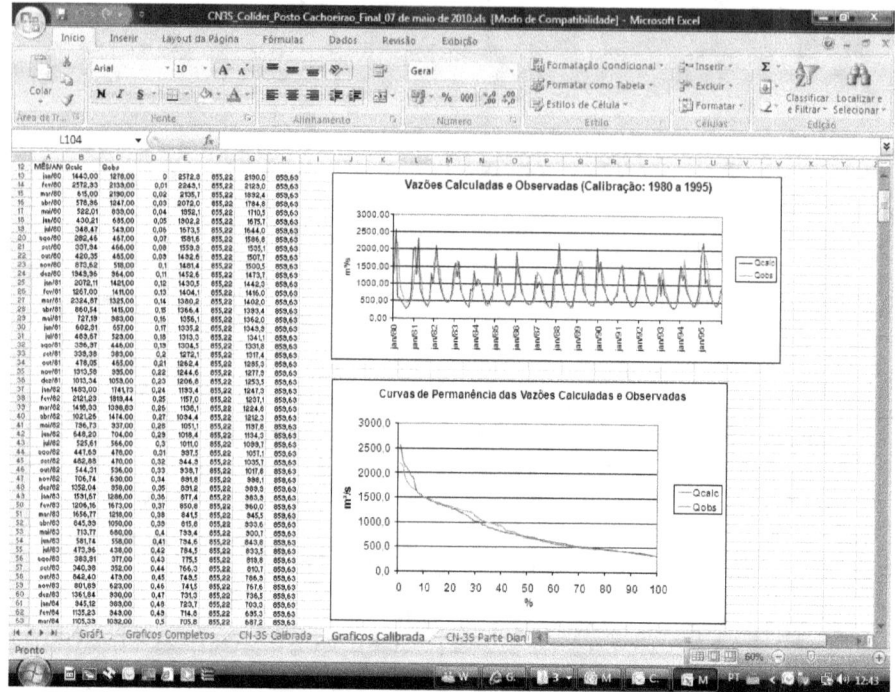

Figure 4. Observed and Calculated streamflow (Calibration Period) and streamflow permanence curves for the Cachoeirão station

Bibliography

BARD, Y. *Nonlinear Parameter Estimation*, Academic, Orlando, Fla., 1974.

BEVEN, K. J. et al. On Hydrological Heterogeneity – Catchment Morphology and Catchment Response, *Journal of Hydrology*, 100, 353-375, 1988.

CANEDO, P. M. Hidrologia Superficial, In: Engenharia Hidrológica, Rio de Janeiro, ABRH, Ed. da UFRJ, vol. 2, 1989.

_____. *The Reliability of Conceptual Catchment Model Calibration*, PhD Thesis, University of Lancaster, U. K., 1979.

CANEDO, P. M., SILVA, L. P. & XAVIER, A. E. *Calibração Automática de Modelos Chuva-Vazão por meio de Técnicas de Suavização*, IV Simpósio Luso-Brasileiro de Hidráulica e Recursos Hídricos, vol. 2, 277-287, Lisboa, 1989.

CLARKE, R. T. Mathematical Models in Hydrology, *Environmental Science Series*, New York, 1973b, 280p.

DELMOMME, J. P. Spatial Variability and Uncertainty in Groundwater Flow Parameters: A Geostatistical Approach, vol. 15, n. 02, 269-280, 1979.

DISKIN, M. H. & SIMON, E. A procedure for the Selection of Objectives Functions for Hydrologic Simulation Models, *Journal of Hydrology*, 34 (1/2), 129-149, 1977.

FLEMING, G. Computer Simulation Techniques in Hydrology, American Alsevier Publ. Co. Inc., N. Y., 1975, 334p.

FREITAS, M.A.S. *Considerações sobre Modelos Determinísticos Chuva Vazão Aplicados às Bacias do Semi-Árido Brasileiro*, Dissertação de Mestrado, UFC, 1991.

FREITAS, M.A.S. Modelos Diários Chuva-Vazão em Bacias do Semi-Árido Brasileiro, *Revista Tecnologia* - UNIFOR, vol. 15, 31-38, 1994.

FREITAS, M.A.S. & A.S. PORTO: Considerações Sobre um Modelo Determinístico Chuva-Vazão Aplicado às Bacias do Semi-Árido Nordestino, *Revista Tecnologia* - UNIFOR, vol. 11, 45-49, 1990.

FREITAS, M.A.S.: Aspectos a Serem Considerados Quando de uma Análise Regional Integrada de Secas, *Revista Tecnologia* - UNIFOR, vol. 17, 9-17, 1996.

FREITAS, M.A.S. Estudos para a Elaboração de Normas de Operação e Manutenção da Barragem Bocaina, SEMAR/SRH/MMA, 2001.

GUPTA, V. K. & SOROOSHIAN, S., Uniqueness and Observability of Conceptual Rainfall-Runoff Model Parameters: The Percolation Process Examined, *Water Resources Research*, vol. 19, nº 01, 267-269, 1983.

KUCZERA, G. Improved Parameter Inference in Catchment Models 1. Evaluating Parameter Uncertain, *Water Resources Research*, vol. 19, n.5, 1151-1162, 1983.

KUSTER, J. L. & MIZE, J. H. *Optimization Techniques with FORTRAN*, McGraw Hill Book Company, 1976.

LINSLEY, R. K. Jr. "Rainfall-Runoff Models – An Overview", Rainfall-Runoff Relationships, edited by V. P. Singh, *Water Resources Publications*, 3-22, 1981.

NOUVELOT, J. F., FERREIRA, P. A. S. & CADIER, E., *Bacia Representativa do Riacho do Navio – Relatório Final*, SUDENE, Série Hidrológica, n. 6, 1979.

O'DONNELL, T. & CANEDO, P. M. The Reliability of Conceptual Basin Model Calibration, Proceedings of the Oxford Symposium on Hydrological Forecasting, IAHS, nº 129, 1980.

PILGRIM, D. H. *et al*. Effects of Catchment Size on Runoff Relationship, *Journal of Hydrology*, 58, 205-211, 1982.

Plano Diretor de Recursos Hídricos da Bacia Hidrográfica dos Rios Canindé/Piauí, no Estado do Piauí, Relatório de Andamento N° 08, Tomo I – Diagnóstico, vol. 08, FAHMA Planejamento e Engenharia Agrícola Ltda, fev. 1999.

Plano Diretor de Recursos Hídricos da Bacia Hidrográfica dos Rios Canindé/Piauí, no Estado do Piauí, Relatório de Andamento N° 08, Tomo I – Diagnóstico, vol. 09, FAHMA Planejamento e Engenharia Agrícola Ltda, fev. 1999.

Plano Diretor de Recursos Hídricos da Bacia Hidrográfica dos Rios Canindé/Piauí, no Estado do Piauí, Relatório de Andamento N° 08, Tomo I – Diagnóstico, vol. 10, FAHMA Planejamento e Engenharia Agrícola Ltda, fev. 1999.

Plano Estadual de Recursos Hídricos, Estudos Hidrológicos, SIRAC/SRH-CE, 1991.

RETNAM, M. T. P. & WILLIAMS, B. J. Input Errors in Rainfall-Runoff Modelling, *Mathematics and Computers in Simulation*, 30, 119-131, Nort-Holland, 1988.

ROSENBROCK, H. H. An Automatic Method for Finding the Greatest or Least Value of a Function, *The Computer Journal*, 3, 175-184, 1960.

SHERMAN, L. W. *Streamflow from Rainfall by the Unit-Graph Method*, Eng. News-Record, 108, 1932.

SMITH, R. E. & HERBERT, R. H. B. A Monte Carlo Analysis of the Hydrologic Effects of Spatial Variability of Infiltration, *Water Resources Research*, 15(2), 419-429, 1979.

SOROOSHIAN, S. & GUPTA, V. K. The Analysis of Structural Identifiability: Theory and Application to Conceptual Rainfall-Runoff Models, *Water Resources Research*, 21(4), 487-495, 1985.

SOROOSHIAN, S., GUPTA, V. K. & FULTON, J. C. Evaluation of Maximum Likelihood Parameter Estimation Technique for Conceptual Rainfall-Runoff Models: Influence of Calibration Data Variability and Length of Model Credibility, *Water Resources Research*, 10(1), 251-259, 1983.

TABORGA, J., FREITAS. M.A.S. *Simulação da Lâmina de Escoamento Mensal*, III Simpósio Luso-Brasileiro de Hidráulica e Recursos Hídricos, VII Simpósio Brasileiro de Hidrologia e Recursos Hídricos, vol. 2, 558-570, Salvador, Bahia, 1987.

TODINI, E. Rainfall-Runoff Modelling – Past, Present and Future, *Journal of Hydrology*, 100, 341-352, 1988.

TUCCI, C. E. M. *et al.* Avaliação da Distribuição Espacial da Infiltração. Revista Brasileira de Engenharia, vol. 6, n° 02, 1988.

TUCCI, C. E. M. *Modelos Hidrológicos*, Editora da Universidade / UFRGS/ABRH, 1998.

VIESMANN, W. *et al.* Introduction to Hydrology, IEP a DUN-Donnelley Publisher, New York, 1977, 704p.

WOOD, E. F. *et al.* Effects of Spatial Variability and Scale with Implications to Hydrologic Modeling, *Journal of Hydrology*, 102, 29-47, 1988.

CHAPTER 13

Floods Prediction in the Amazon River Basin – Brazil

Marcos Airton de Sousa Freitas

Specialist in Water Resources at the National Water Agency - ANA; Ministry of
Regional Development - MDR; masfreitas@ana.gov.br

Abstract: This paper focuses the problem of flood modelling for the Amazonian
hydraulic basins, in special, forecast for the city of Manaus. This makes an
analysis of historical floods occurred in Manaus, as well as it presents flood
forecast models for this city. Linear and not linear regressions models, as well
as artificial neural network model for flood forecast have been presented. With
those models it is possible to prognose a flood in Manaus station with one
month of antecedence.

Keywords: flood prediction, amazon watershed, artificial neural network

1. Introduction

The Legal Amazon, a concept created in the 1950s as a strategy to stimulate
national development and the occupation of Brazilian territory, has an
extension of just over 5 million square kilometers. It corresponds to 60% of the
national territory and is distributed in nine states. Brazil holds 67.7% of the
Continental Amazon - which includes areas from Bolivia, Colombia, Guyana,
French Guiana, Ecuador, Peru, Suriname and Venezuela. Located in the North of
the country, the Brazilian Amazon borders with all these South American
states, with exception of Ecuador, totaling a border strip of 12 thousand
kilometers (ACTO, 2004).

The Amazon is known worldwide for the availability of water and the diversity
of ecosystems, such as dry land forests, flooded forests, floodplains, igapós,
open and closed fields. The Amazon is home to a multitude of plant and animal
species, with about 1.5 million cataloged plant species, 3,000 species of fish,
950 types of birds, as well as insects, reptiles, amphibians and mammals.

The most extensive hydrographic network on the globe, the Amazon Region
occupies a total area of 6,925,674 km^2, from its sources in the Peruvian Andes
to its mouth in the Atlantic Ocean, in the north of Brazil, covering the territories
of Brazil, Colombia, Bolivia, Ecuador , Guyana, Peru and Venezuela.

The estimated long-term average flow of the Amazon River is of the order of 133,861 m^3 / s (68% of the country's total). The contribution of foreign territories to the flow of the hydrographic region is 71,527 m^3 / s. The greatest demands for water use in the region occur in the sub-basins of the Madeira, Tapajós and Negro rivers, and correspond to the use for irrigation (37% of the total demand). Urban Demand represents 17% of the region's demand (10.9 m^3 / s). In general, the estimated consumption is insignificant when compared to water availability.

One of the characteristics of this region has to do with the observed deforestation. Until January 1978, the area deforested in the states inserted in the Amazon Region corresponded to 85,100 km^2, the result of human actions in the basin over more than four centuries. From the 70s onwards, there was a significant increase in the occupation of the region, resulting in the expansion of deforested areas as result of this dynamic. In 1999, there was a deforested area of 440,630 km2. National Institute for Space Research (INPE) data indicate, for the years 1999 and 2000, rates of gross deforestation of 17,259 and 19,836 km^2 / year, respectively (http://www.ana.gov.br/mapainicial/pgMapaA.asp).

The floods in the Amazon Basin are the result of a natural phenomenon, as part of the dynamics of the river. In the specific case of the city of Manaus and its surroundings, they mainly result from the affluence of the Solimões and Negro rivers. Due to the wide extension of the basin, the predictability of floods in the city of Manaus, the main municipality in the basin, is relatively easy and can be implemented several days in advance.

2. Description of the Models

CPRM - Geological Service of Brazil has been developing, since 1989, the Project called Manaus Flood Alert, where the annual flood process is monitored in the Solimões / Amazonas / Negro system. A 75, 45 and 15 antecedent days predictions are made for a maximum flood forecast for the Negro River, in the Port of Manaus station (Figure 1).

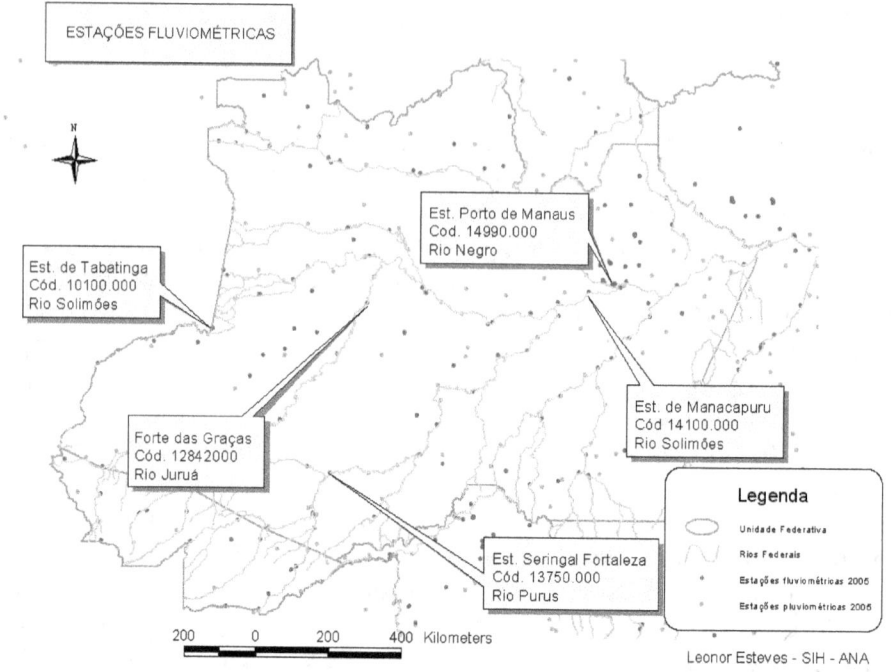

Leonor Esteves - SIH - ANA

Figure 1: Location of the used fluviometric stations.

Given the availability of data from the Porto de Manaus station, having water level values from September 1902, it was possible to establish regressions between peak flood and water level values 75, 45 and 15 days in advance (HIDRO-ANA, 2002). Water levels from March 30, April 30 and May 31 were used to estimate the water level on June 15.

For this purpose, in 2004, CPRM used the following equations:

- March 31 forecast: $H_{max} = 1627.23175 \ln H_{31\ March} - 9925.32303$, with correlation coefficient, $R = 0.84273$;
- April 30 forecast: $H_{max} = 8.2156\ H_{30\ April}{}^{\wedge} 0.73998$, with correlation coefficient $R = 0.93740$;
- May 31 forecast: $H_{max} = 2.199925\ H_{31\ May}{}^{\wedge} 0.90217$, with correlation coefficient $R = 0.97659$.

In the three cases H_{max} is the maximum height of the flood; $H_{31\ March}$, $H_{30\ April}$, $H_{31\ May}$ are the observed water levels at the Port of Manaus station, on March 31, April 30 and May 31, respectively, all in centimeters.

For 2005, using the CPRM regression equations, we have the following values for the Porto de Manaus post: $H_{31 \text{ March}} = 2601 \rightarrow H_{max} = 2871$; $H_{30 \text{ April}} = 2716 \rightarrow H_{max} = 2855$ (according to FAX nº 157 / SUREG-MA / 2005) and $H_{31 \text{ May}} = 2808 \rightarrow H_{max} = 2840$. This concept is based on the observation of the typical diagram of the fluviometric station in Porto de Manaus (Roadway), whose historical series started in September 1902, from which the correlations between the quotas of a given day with the quota of the peak of the flood were inferred (Boletim CPRM, 2005).

Based on this principle, linear correlations between the water levels were determined initially on March 31, April 30 and May 31, for the Manaus post, and the water level on June 15 for the same station, using the data from 1902 to 2003. From Figure 2 to Figure 4 linear regression equations found are shown. The water level values found were: $H_{31 \text{ March}} = 2601 \rightarrow H_{max} = 2864$; $H_{30 \text{ April}} = 2716 \rightarrow H_{max} = 2845$ and $H_{31 \text{ May}} = 2808 \rightarrow H_{max} = 2833$.

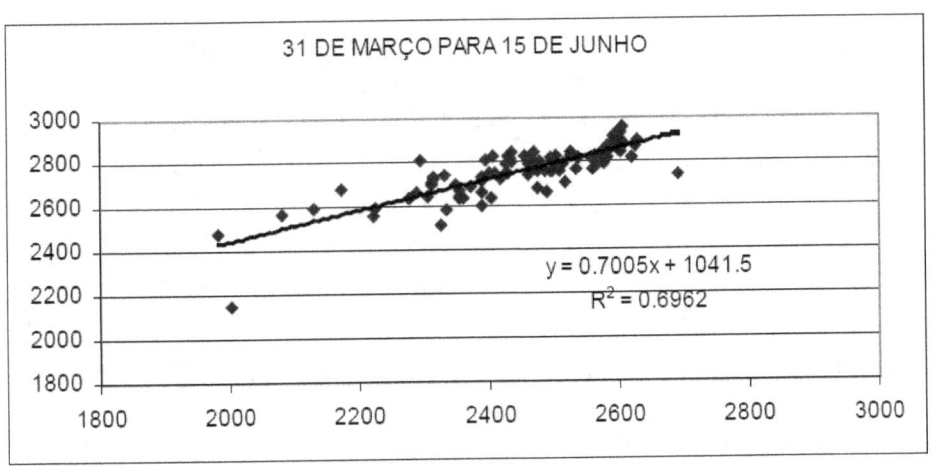

Figure 2: Linear regression equation for prediction on March 31 to June 15.

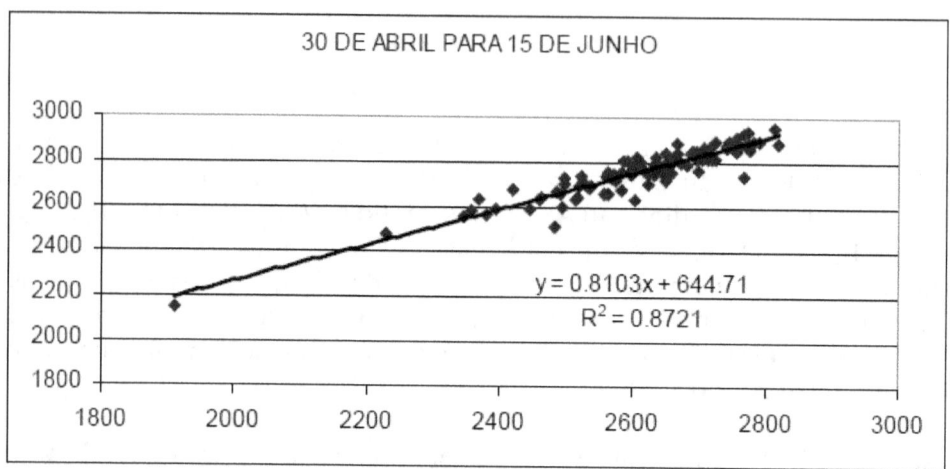

Figure 3: Linear regression equation for prediction on April 30th to June 15th.

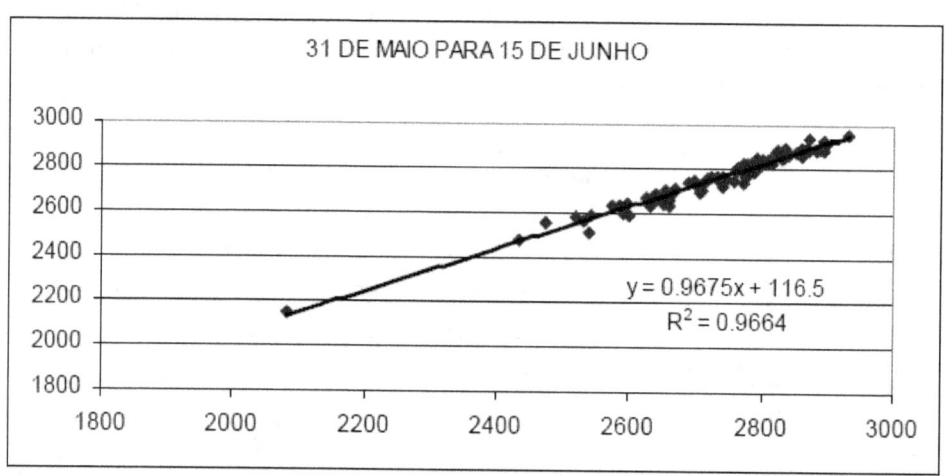

Figure 4: Linear regression equation for prediction on May 31 to June 15.

For each of the three dates, the best-fit curve between the water levels on March 31, April 30 and May 31 was determined for the Manaus station, and the water level on June 15 for the same station, using data from 1902 to 2003 are shown. Figures 5 to 7 show the curves found. The water level values found were: $H_{31\ March} = 2601 \rightarrow H_{max} = 2860$; $H_{30\ April} = 2716 \rightarrow H_{max} = 2842$ and $H_{31\ May} = 2808 \rightarrow H_{max} = 2834$.

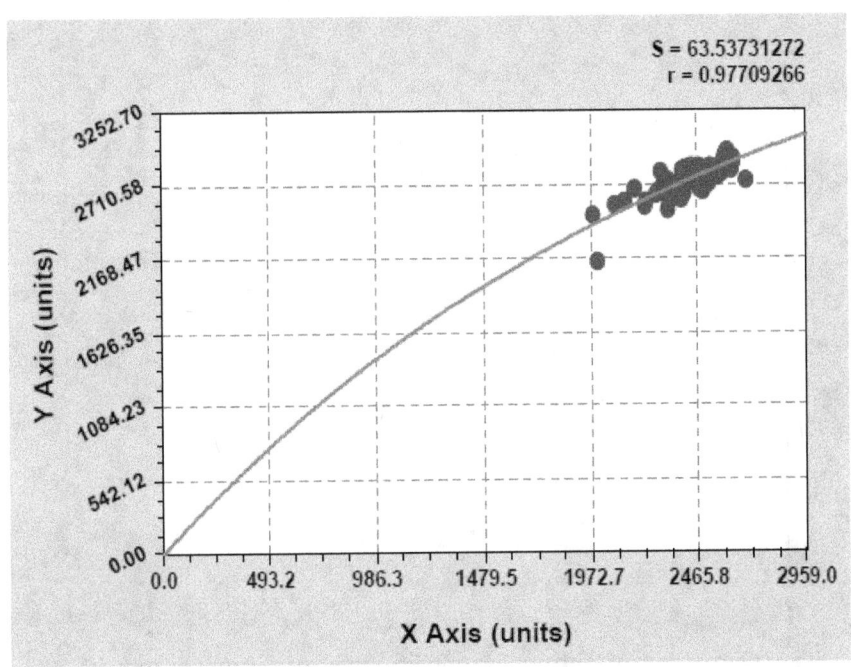

Figure 5: Regression equation for prediction on March 31 to June 15.

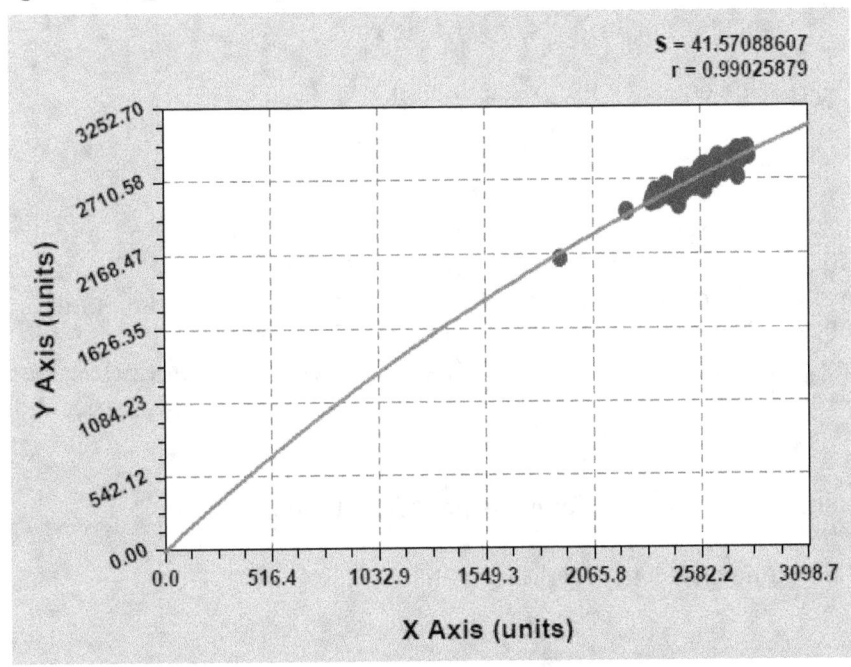

Figure 6: Regression equation for prediction on April 30th to June 15th.

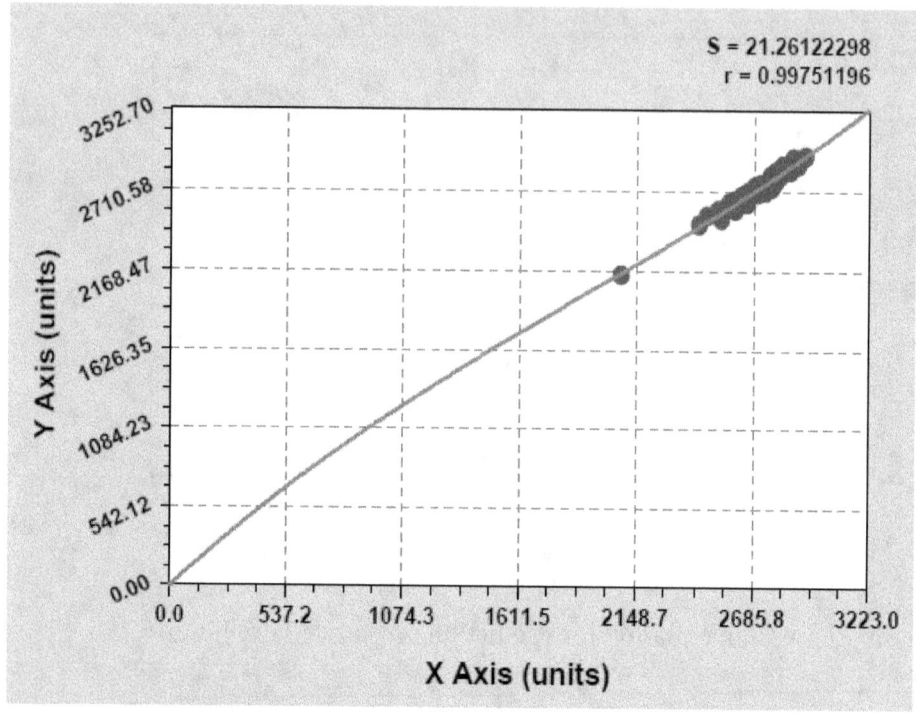

Figure 7: Regression equation for prediction on May 31 to June 15.

The equations found were as follows:

- March 31 forecast: $H_{max} = a(1-e^{bx})$, with a = 4536.8463 and b = 0.00038268861; correlation coefficient, R = 0.97709266;
- April 30 forecast: $H_{max} = a(1-e^{bx})$, with a = 5993.1868 and b = 0.00023665087; with correlation coefficient R = 0.99025879;
- May 31 forecast: $H_{max} = a + bx + cx^2 + dx^3$, with a = -0.049291611; b = 1.3914972; c = -0.00027202643; d = 4.8393553E-8; with correlation coefficient R = 0.99751196.

Table 1 summarizes the results found using this methodology.

Table 1: Maximum water level prediction.

Cota	CPRM	ANA_Simples	ANA_Outras Curvas
31 de março: 2601	2871	2864	2860
30 de abril: 2716	2855	2845	2842
31 de maio: 2808	2840	2833	2834

Models based on Artificial Neural Networks

Models based on artificial neural networks were also used. Significant progress has recently been made in the fields of Pattern Recognition and Systems Theory with the use of Artificial Neural Networks (ANN). Neural networks are flexible mathematical structures capable of identifying non-linear relationships and describing complex processes. The name Artificial Neural Networks is given to models, which attempt to reproduce the structure and functioning of neurons in the brain (KOSKO, 1992; FREITAS and BILLIB, 1997). An ANN consists of a many number of elements, called neurons (cells, units) and a many number of connections, known as synapses (Figure 11). Each link (links, connections) is associated with a weight, which is intrinsically related to the network's learning capacity (FREITAS, 1998a).

The creation of an ANN is based on the following tasks (FREITAS, 1997):

- Determination of the characteristics of the network: topology; types of connections; ordering of connections and weights;
- Determination of the characteristics of neurons: input, activation or transfer and output functions;
- Determination of the network dynamics: generation of the initial values of the connection weights; optimization processes and learning rules.

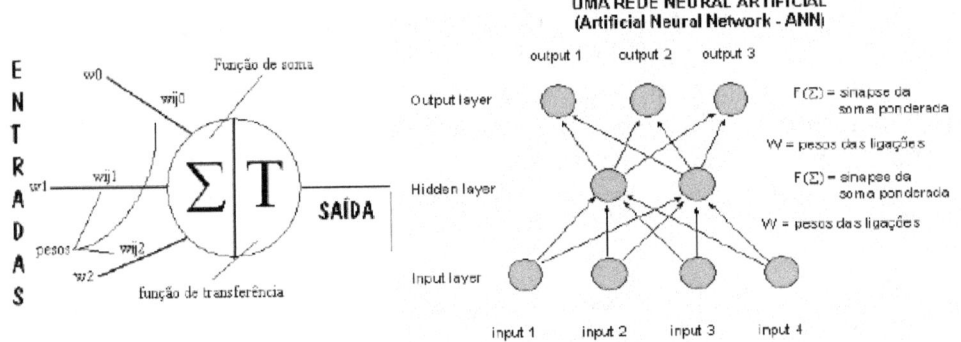

Figure 11: Representation of an Artificial Neural Network.

Depending on the dynamics of the network, as well as neurons and topology, an infinite number of Artificial Neural Networks was developed, namely: the Perceptron model (ROSENBLATT, 1962); the Backpropagation model (RUMMELHART et al., 1986); the ADALINE model (WIDROW and HOFF, 1960); the KOHONEN network (KOHONEN, 1984); the HOPFIELD model (HOPFIELD, 1982), the ART model (CARPENTER and GROSSBERG, 1987) etc. Among these, the Backpropagation method (feedback) is the most known and used method. A simple learning function is given only with the learning parameter.

The backpropagation algorithm seeks to minimize the error obtained by the network by adjusting weights and thresholds so that they correspond to the coordinates of the lowest points on the error surface. For this he uses a gradient method (CARVALHO et al., 1998). Modeling takes place, therefore, through the training, validation and forecasting phases.

Applications of neural networks in water resources area can be seen in ALENCAR et al. (1998), BARROS & FREITAS (1998), FREITAS (1998b), as well as RIBEIRO & FREITAS (1998).

One of the optimization software for this type of network is Slug3 (developed in Pascal language) with three layers of neurons where you can implement the number of neurons per layer (topology), as well as modify the learning rate, the term 'momentum' , the number of cycles required for adjustment and the margin of error that you want to obtain.

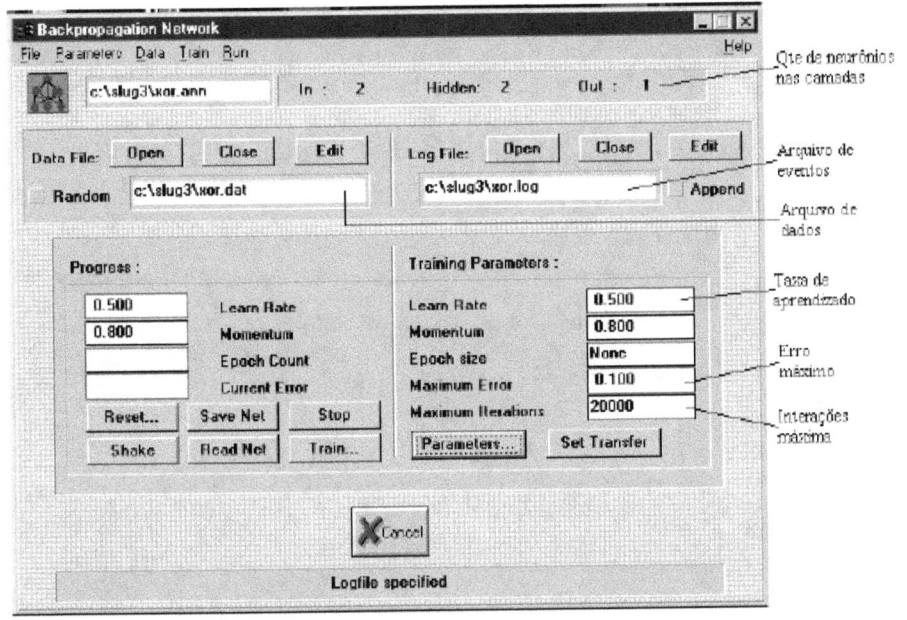

Figure 12: The Artificial Neural Network Slug3 software

The functioning of a network is based on calculations of weights between neurons, using real data (input data and the respective results). The network 'learns' for example the (non-linear) relationship between the data. Normally, the criteria for stopping the training process are normally the termination after a certain number of cycles or the termination after reaching a pre-fixed mean square error.

With water levels on March 31, April 15 and April 30 as input values for the Manaus station, a neural network was adjusted, with the water level as an output on June 15. The series from 1903 to 1977 was used for model calibration (Figure 13) and the period from 1978 to 2003 for validation (Figure 14). The water level value forecast for June 15, 2005, with H $_{31\ March}$ = 2601; H $_{April\ 15}$ = 2645 and H $_{April\ 30}$ = 2716 → H $_{max}$ = 2860. Similarly, by adjusting another neural network, the water level predicted for July 1, 2005, with H $_{15\ April}$ = 2645; H $_{30\ April}$ = 2716 and H $_{15\ May}$ = 2767 → H $_{max}$ = 2840. In the latter case, the series from 1903 to 1977 was also used for the calibration of the model (Figure 15) and the period from 1978 to 2003 for validation (Figure 16).

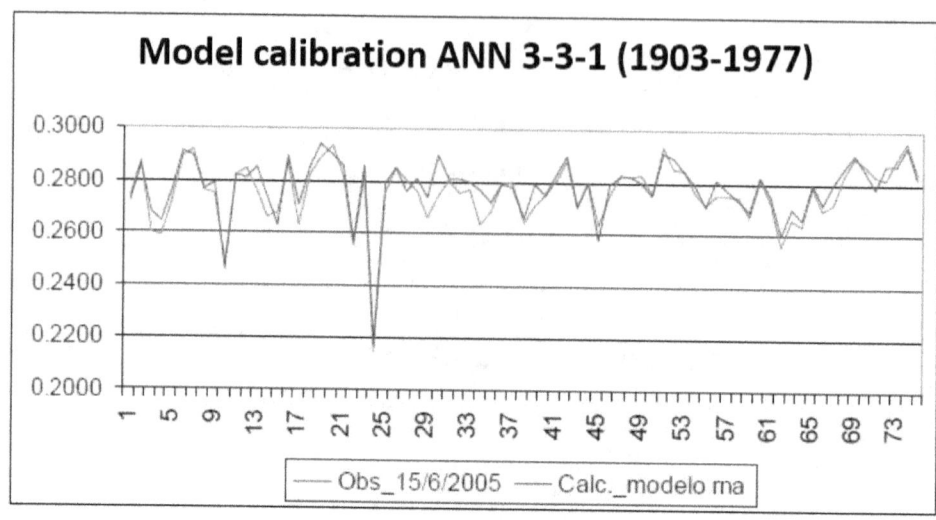

Figure 13: Calibration of an ANN for the Manaus station (June 15, 2005).

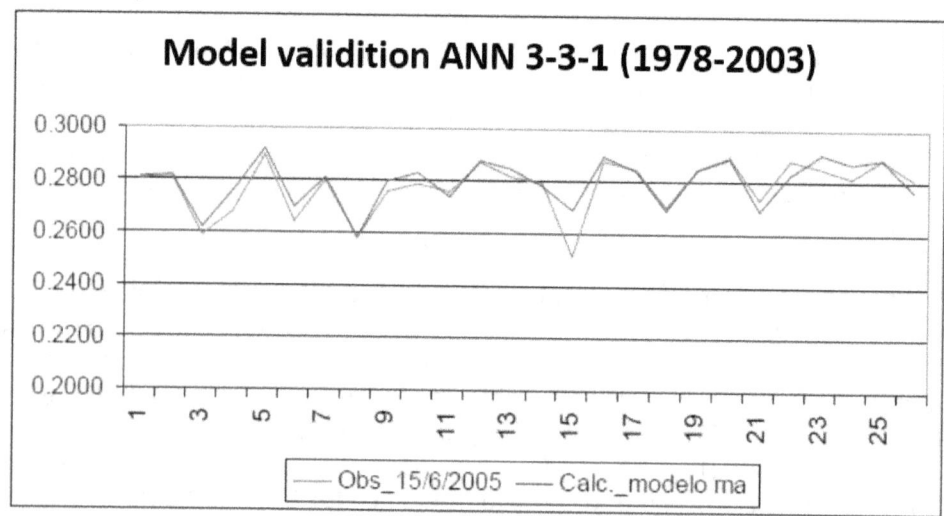

Figure 14: Validation of an ANN for the Manaus station (June 15, 2005).

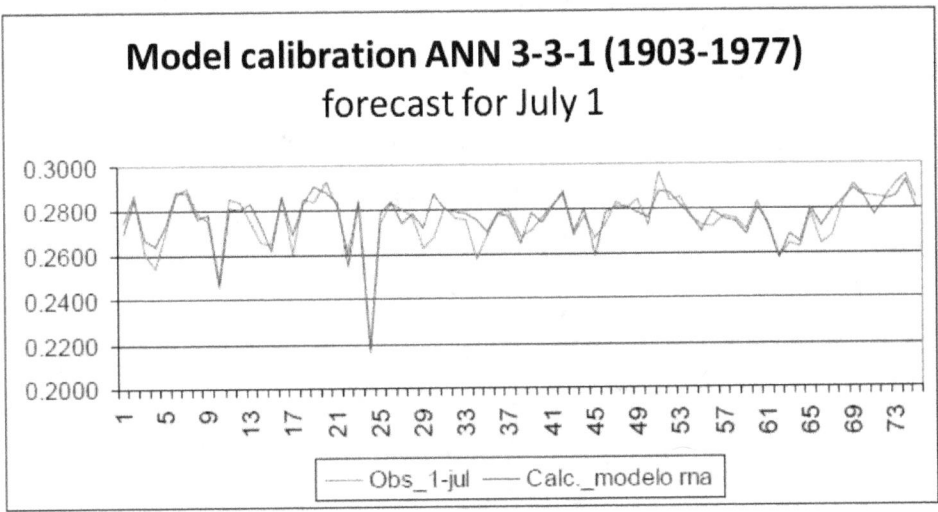

Figure 15: Calibration of an ANN for the Manaus station (July 1, 2005).

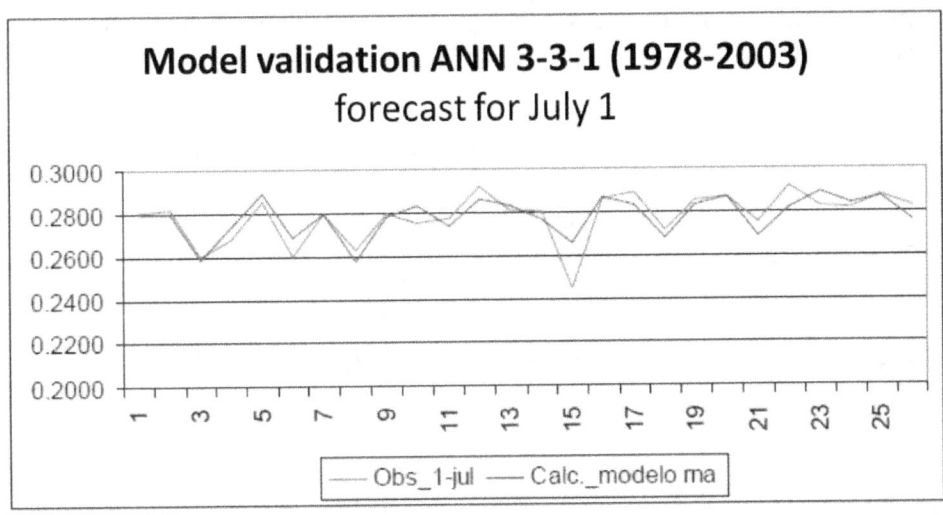

Figure 16: Validation of an ANN for the Manaus station (July 1, 2005).

With water levels as of April 30, May 15 and May 31 as input values for the Manaus station, a neural network was adjusted, with the water level as an output on June 15. The series from 1903 to 1977 was used for model calibration (Figure 17) and the period from 1978 to 2003 for validation (Figure 18). The water level value forecast for June 15, 2005, with $H_{April\ 30} = 2717$; H_{May}

$_{15}$ = 2785 and H $_{May\ 31}$ = 2808 → H $_{max}$ = 2810. The water level observed on June 15 was exactly 2810.

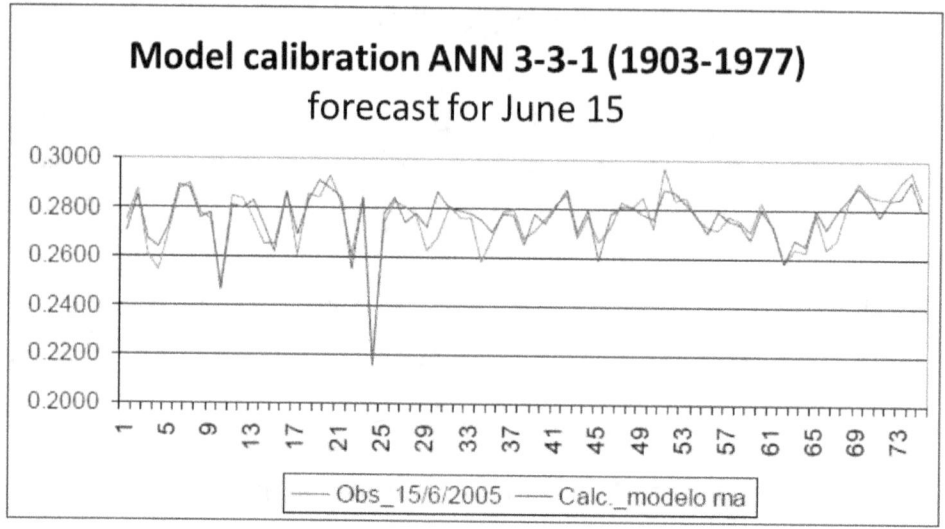

Figure 17: Calibration of an ANN for the Manaus station (June 1, 2005), with water levels on April 30, May 15 and May 31.

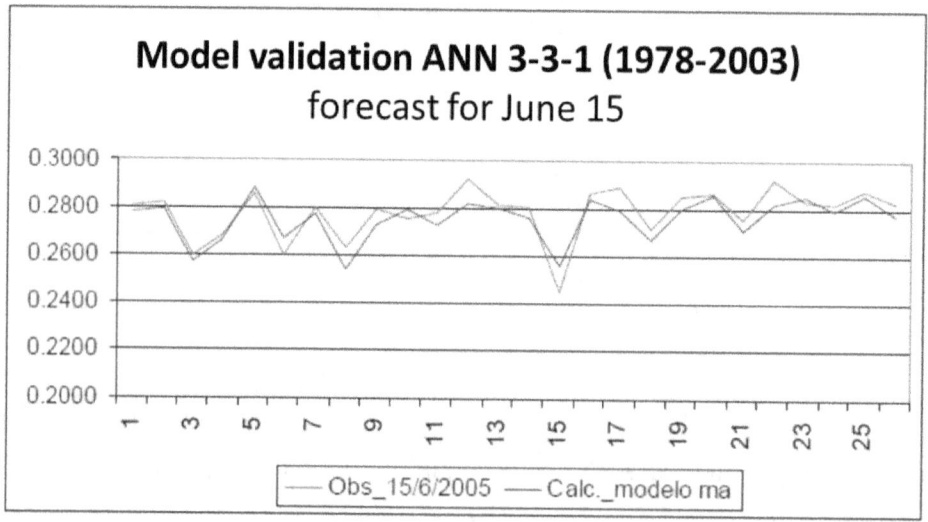

Figure 18: Validation of an ANN for the Manaus station (June 1, 2005), with water levels on April 30, May 15 and May 31.

Regressions were also established with stations upstream of Manaus, namely the river stations of Tabatinga, Seringal Fortaleza and Forte das Garças, aiming to predict the values of water levels in the river stations of Manacapuru and Porto de Manaus.

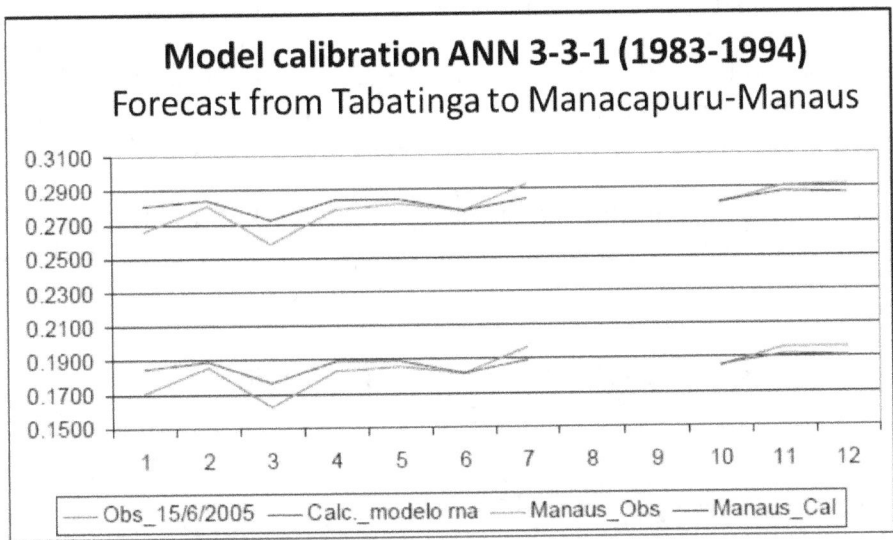

Figure 19: Calibration of an ANN for the Manaus station (June 15, 2005) using data from the Tabatinga station.

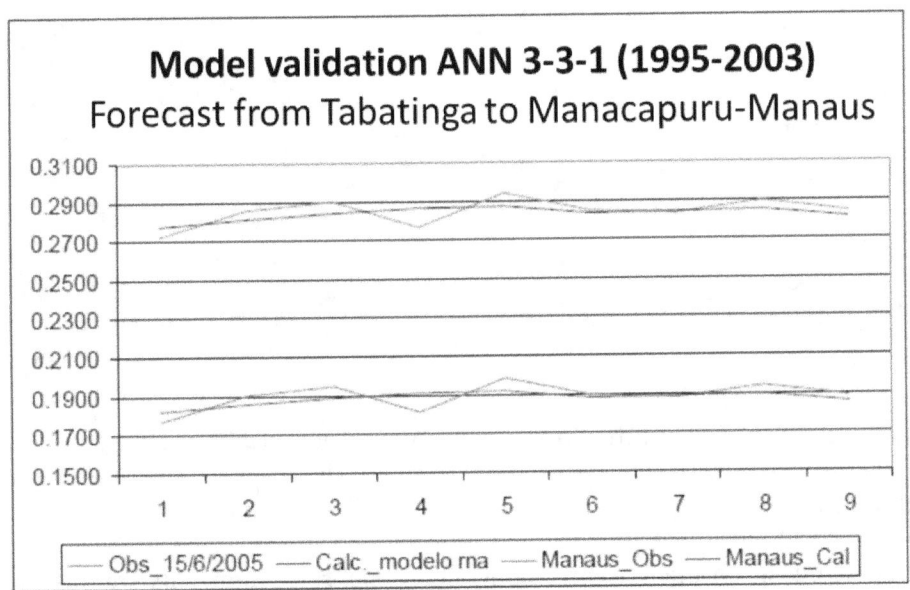

Figure 20: Validation of an ANN for the Manaus station (June 15, 2005) using data from the Tabatinga station.

Using as input values the water levels on April 15, April 30 and May 15, for the Tabatinga station, a neural network was adjusted, with the water level as the outlet, for the Manacapuru station (and for correlation to the Manaus station), on June 15. The series from 1983 to 1994 was used for model calibration (Figure 19) and the period from 1995 to 2003 for validation (Figure 20). The water level value forecast for June 15, 2005, with the water levels, in Tabatinga, from $H_{15\,April} = 1066$; $H_{30\,April} = 1154$ and $H_{15\,May} = 1139 \rightarrow H_{max} = 2744$.

The water levels of the highest floods for the Porto de Manaus station, from 1902 to 2004, as well as the corresponding water levels for the year 2005, are shown in Figure 21.

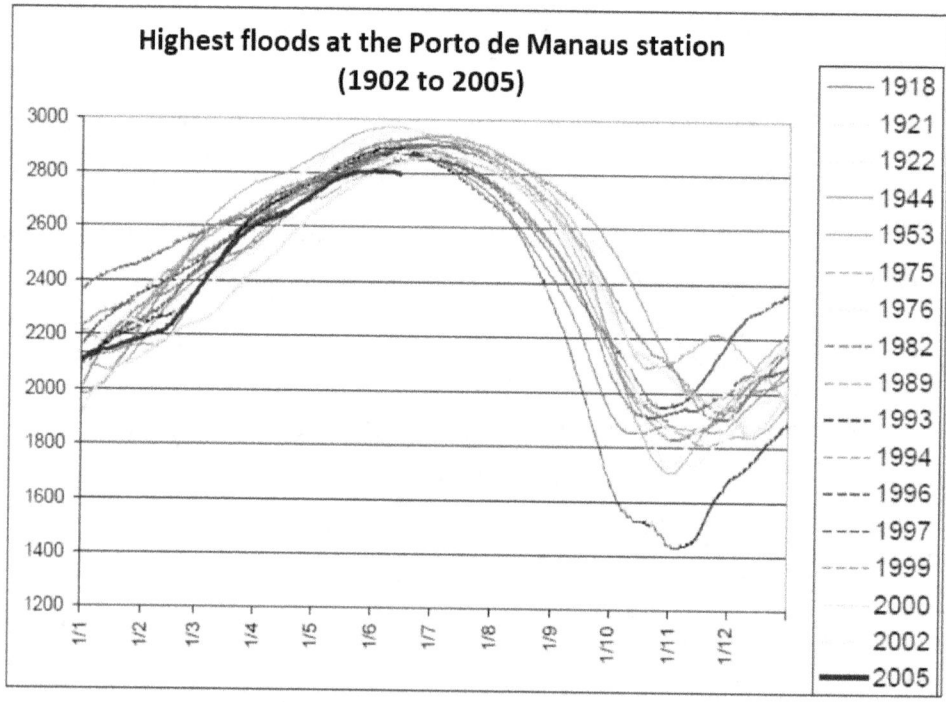

Figure 21: Highest floods at the Porto de Manaus station (1902 to 2005).

3. Results

The models based on linear regressions, nonlinear regressions, as well as the artificial neural network models showed satisfactory flood forecasting for the city of Manaus up to 75 days in advance. In the case of neural network models, the period from 1903 to 1977 was used for the calibration of the model and the period from 1978 to 2003 for validation. Forecasts were made, using the different models, for the year 2005.

4. Conclusions and Recommendations

The implementation of regression models (linear and non-linear) and models based on artificial neural networks proved to be satisfactory in flood forecasting in the city of Manaus. In the future, the intention is to incorporate data from other fluviometric and rainfall stations to have more reliable forecasts, not only for the city of Manaus, but for the strategic points of the Amazon basin.

The regression models predicted 75 days in advance, for the Porto de Manaus station, that the water level should reach, on June 15, values varying from 28.71m to 28.60m and on July 1 water level ranging from 28.59m to 28.56m.

With 45 days in advance, the models predicted for the Port of Manaus station that the water level should reach, on June 15, the water level ranging from 28.55m to 28.42m and on July 1, the level ranging from 28.46m to 28.40m.

With 30 days in advance, the models predicted for the Port of Manaus station, that the water level should reach, on June 15th, the water level varying from 28.40m to 28.33m and on July 1st, the level of water ranging from 28.32m to 28.29m. The emergency water level adopted by CPRM is 28.50m.

These values are provisional and can be improved as new data is incorporated into the models. Forecasts from Tabatinga, Seringal de Fortaleza and Forte das Garças stations were hampered by the lack of access to real-time information.

The models of artificial neural networks, having as input values the water levels on April 30, May 15 and May 31, for the Manaus station, managed to accurately predict the flood value on June 15, 2005. These models are being improved for day-to-day forecasting, in other parts of the basin, incorporating other variables, such as precipitation.

5. Bibliographic References

ALENCAR, P. F.; RIBEIRO, A V. R.; FREITAS, M. A. S. (1997). "Modelos Computacionais Baseados na Biologia: Algoritmos Genéticos e Redes Neurais Artificiais", Revista Tecnologia - UNIFOR, vol. 18, 92-98.

BARROS, F. R. M.; FREITAS, M.A.S. (1998). "Alerta de Cheias por Redes Neurais Artificiais", in Anais do IV Encontro de Iniciação à Pesquisa, Resumos, Universidade de Fortaleza, Ceará.

CARPENTER, G. A.; GROSSBERG, S. (1987). "Massively Parallel Architecture for a Self-Organizing Neural Pattern Recognition Machine", Computer Vision, Graphics, and Image Processing, vol. 37, 54-115.

CARVALHO, A. C. P. L. F.; BRAGA, A. P. L.; LUDEMIR, T.B. (1998). Fundamentos de Redes Neurais Artificiais. 11ª Escola de Computação, Rio de Janeiro, DCC/IM, COPPE/Sistemas, NCE/UFRJ, 246p.

FREITAS, M.A.S. (1997). "O Fenômeno do El Niño e as Secas no Ceará: A Previsão através de Modelos Estatísticos e de Redes Neurais Artificiais", in Anais do I Fórum Interamericano de Gestão de Recursos Hídricos, de 10 a 14 de novembro de 1997, Fortaleza - CE (http://www.iica.org.br).

FREITAS, M.A.S. (1998). Neurocomputação Aplicada. Gráfica UFPI, 1998, 48p.

FREITAS, M.A.S. (1998) "A Decision Support System for Drought Forecasting and Reservoirs Management in Northeast Brazil", in Anais do VIII Congresso Latino-Americano e Ibérico de Meteorologia, X Congresso Brasileiro de Meteorologia, Brasília - DF.

FREITAS, M.A.S.; BILLIB, M. H. A. (1997). "Drought Prediction and Characteristic Analysis in Semi-arid Ceará - Northeast Brazil", in Anais do Symposium "Sustainability of Water Resources Under Increasing Uncertainty", IAHS Publ. Nº 240, pág. 105-112, Rabat, Marrocos.

HIDRO-ANA (2002). Sistema de Informações Hidrológicas, Versão 1.0.8, Compilação 1.0.8.2508.

HOPFIELD, J. J. (1982). "Neural Networks and Physical Systems with Emergent Collective Computational Abilities", Proc. of the National Academy of Science, USA, Biophysics, 79, 2554-2558.

KOHONEN, T. (1984). Self-Organization and Associative Memory, Springer Verlag, New York.

KOSKO, B. (1992). Neural Networks and Fuzzy Systems, Prentice Hall.

NOTA TÉCNICA nº 027/2005/SIH-ANA: Estações pluviométricas e fluviométricas localizadas à montante de Manaus.

ORGANIZAÇÃO DO TRATADO DE COOPERAÇÃO AMAZÔNICA – Boletim OTCA, Ano I, Nº 3 – Dezembro 2004 / Fevereiro 2005.

RIBEIRO, A.V. R.; FREITAS, M.A.S. (1998). "Aplicações dos Algoritmos Back_Prop_Momentum e Back_Prop_Through_Time em Recursos Hídricos", in Anais do IV Encontro de Iniciação à Pesquisa, Resumos, Universidade de Fortaleza, Ceará.

ROSENBLATT, F. (1962). Principles of Neurodynamics, Spartan Books, New York.

RUMMELHART, D. E.; HINTON, G. E.; WILLIAMS, R. J. (1986). "Learning Representations by Back-propagations Errors", Nature, 323(9), 533-536.

WIDROW, B.; HOFF, M. E. (1960). "Adaptive Switching Circuits", IRE WESCON Convention Record, New York, 96-104.

PART III – SUSTAINABILITY IN WATER RESOURCES

"Não sonhe pequeno, pois esses sonhos não têm poder para mover os corações dos homens". Johann Wolfgang von Goethe.

"Do not dream small, because these dreams have no power to move the hearts of men". Johann Wolfgang von Goethe.

CHAPTER 14

The Water-Climate-Forest Nexus: Integration of Policies, Governance and Social Participation in Brazil

Marcos Airton de Sousa Freitas[1]

[1] Specialist in Water Resources at the National Water Agency - ANA; Ministry of Regional Development - MDR; masfreitas@ana.gov.br

Abstract - This article deals with the discussion of the integration and implementation of three important Public Policies in Brazil, namely: the National Water Resources Policy, the National Policy on Climate Change and the Forest Policy. The National Water Resources Policy, created by Law 9.433 / 1997, has been implemented throughout Brazil, through its instruments, such as the grant, collection, basin plans, etc. The most recent National Policy on Climate Change has numerous instruments, such as the National Plan and the National Fund on Climate Change. Promotion issues are also addressed. Until the Federal Constitution, in 1988, Brazilian forests had not yet received specific treatment, at the constitutional level, either as environmental heritage or as forest heritage, despite the Forest Code, published in 1965, which remains the most important with regard to the protection of forests in Brazil, currently the stage for the possibility of changes and intense discussions. Recently, new forest policies have been established: National Forest Program (2000), and Public Forest Management Law (2006). The question of the role of the State, the market and the participation of society, both within the scope of the basin committees (governance of water resources) and in forest management (forest governance), as well as in mitigation and adaptation to climate change are also highlighted. Such aspects permeate the discussion about the themes, trying to point out the main difficulties, obstacles and advances. Analyzing them is, therefore, fundamental to continue in the effective transforming potential of these policies in the face of the new challenges imposed.

Keywords: Water-Climate-Forest Nexus, Integration of Policies, Governance

1. Introduction

In the last decades, several studies have analyzed water resource management and the National Water Resources Policy in the light of economic, environmental, ethical and social aspects (Boff, 2003; Lobato da Costa, 2003; Porto & Lobato da Costa, 2004; Freitas, 2009; Freitas, 2010, etc.). However, the current management situation points to a crisis scenario (Haddad, 2008). The crisis around water reflects the crisis of conscience of our civilization and of the current, unequal, exclusive and depleting and exhausting world model of natural resources. In the process of building the sustainable management model of water resources in vogue, the big challenge is to establish a shared and decentralized power relationship, creating an opportunity for social participation, building consensus, settling conflicts and agreeing on unity in diversity (MMA, 2008).

Some experiences in several hydrographic basins, carried out through the implementation of the instruments of the National Water Resources Policy - PNRH (water resource rights grant, hydrographic basin master plans, charging for the use of water resources, framing of water bodies according to predominant use, etc.), as well as inspection campaigns, training programs, financing and stimulating research, to the community organization have been adopted in Brazil. However, little has been researched to verify the real impacts and effects of these plans, programs and projects, as well as the integration of this policy with climate change and forest management policies.

2. National Water Resource Policy - PNRH

Observing the history of the management of water resources in Brazil, it goes back to several institutions, in most cases, responsible sometimes for some user sectors (hydro-energetic generation, sanitation, irrigation, etc.), sometimes with operations restricted to a specific region, for example of the National Department for Works Against Drought - DNOCS (created in 1909).

The increase in demand, coupled with the scarcity and deterioration of the quality of water resources, cause serious conflicts to the multiple use of water, requiring new management paradigms. The Federal Law nº 9.433 / 1997, which instituted the PNRH, has as basic principles, among others, the recognition of water as a vulnerable, finite resource with economic value and the decentralized and participative management of water resources.

The water management models, that is, the institutional (legal and organizational) and financial mechanisms, have evolved over time in three distinct phases, namely: i) the bureaucratic model; ii) the economic-financial model; and iii) the systemic model of participatory integration. Such models have a close relationship with the models of organization management and with the concept of society of Habermas' theory of modernity (Freitas, 2009).

The management of water resources is a complex task and involves several conflicting interests. Thus, the public power, without giving up its role as a managing and coordinating body, recognizes the need to promote a decentralization of management, allowing the intervention of representatives of the various segments involved. This happens through social negotiation and formation of the Hydrographic Basin Committees. The legal instruments for implementing this model are, among others, the following:

1. River Basin Plans aim to support and guide the implementation of the Water Resources Policy and its management, which are prepared by river basin, by State and for the country.

2. Granting the right to use water resources, which is the administrative act by which the granting public authority allows the recipient to use water resources, for a specified period, under the terms and conditions expressed in the respective act.

3. Classification of bodies of water in classes, which aims to ensure the quality of water compatible with the most demanding uses.

4. Charging for the use of water resources, standing out as an economic-financial instrument, aims to: i) recognize water as an economic good; ii) obtain financial resources to finance the programs and interventions contemplated in the plans.

5. Water Resources Information System is a system for collecting, treating, storing and retrieving information about water resources and factors involved in their management.

The aspect of integrated public management deserves special attention, as it constitutes an instrument of institutional framework for conflicts, inevitable in a continental country with enormous diversity. It is a concept that goes back to the countless social movements, since the 1970s, they are part of the Brazilian political reality. Thus, the Federal Constitution of 1988 provided for the organization of the National Water Resources Management System - SINGREH, formed by a set of legal and administrative mechanisms in order to coordinate the integrated management of water resources.

The National Water Agency - ANA is an autarchy under a special regime, created by Law No. 9.984 / 2000, endowed with administrative and financial autonomy, linked to the Ministry of the Environment, part of the National

Water Resources Management System and aims to implement, in its sphere of duties, the PNRH.

Currently, there are 7 Basin Committees installed in rivers in the Union. The Water Agencies are technical executive entities that will act in support of the executive secretariat of the basin committees.

Thus, in summary, SINGREH as participating bodies: i) the National Water Resources Council; ii) ANA; iii) the Water Resources Councils of the States and the Federal District; iv) the Basin Committees; v) the federal, state, Federal District and municipal government agencies whose competences are related to the management of water resources; vi) Water Agencies.

3. National Climate Change Policy - PNMC

Law 12,187 / 2009 instituted the National Policy on Climate Change - PNMC. In accordance with its art. 3rd, the PNMC and the resulting actions, carried out under the responsibility of political entities and public administration bodies, will observe the principles of precaution, prevention, citizen participation, sustainable development and that of common, but differentiated responsibilities, this internationally.

Regarding the measures to be adopted in its execution, the following will be considered: i) everyone has a duty to act, for the benefit of present and future generations, to reduce the impacts resulting from anthropic interference on the climate system; ii) measures will be taken to predict, avoid or minimize the identified causes of climate change with anthropic origin in the national territory, on which there is reasonable consensus on the part of the scientific and technical means engaged in the study of the phenomena involved; iii) actions at the national level to face climate change, current, present and future, must consider and integrate the actions promoted at the state and municipal levels by public and private entities.

Among the numerous PNMC guidelines, the following can be highlighted: i) the commitments assumed by Brazil in the Framework Convention, in the Kyoto Protocol and in the other documents on climate change to which it will become a signatory; ii) actions to mitigate climate change in line with sustainable development; iii) adaptation measures to reduce the adverse effects of climate change and the vulnerability of the environmental, social and economic systems; iv) integrated strategies for mitigating and adapting to climate change; v) stimulating and supporting the participation of the federal, state, district and municipal governments, as well as the productive sector, the academic environment and organized civil society, in the development and execution of policies, plans, programs and actions related to change of the climate.

However, it is worth highlighting some of the instruments of the National Policy on Climate Change, namely: i) National Plan on Climate Change; ii) the National Fund on Climate Change; iii) Action Plans for the Prevention and Control of Deforestation; iv) Brazil's National Communication to the Framework Convention, in accordance with the criteria established by that Convention and its Conferences of the Parties; v) the resolutions of the Interministerial Commission on Global Climate Change; etc.

With regard to the institutional instruments for the performance of the National Climate Change Policy, the following are included: i) the Interministerial Committee on Climate Change; ii) the Interministerial Commission on Global Climate Change; iii) the Brazilian Forum on Climate Change; iv) the Brazilian Research Network - Rede Clima; v) the Meteorology, Climatology and Hydrology Activities Coordination Commission.

According to Art. 8, the official financial institutions will provide specific lines of credit and financing to develop actions and activities aimed at inducing the conduct of private agents to observe and execute the PNMC, within the scope of their social actions and responsibilities.

4. Forest Policy

The first Forest Code was established by Decree 23.793 / 1934, which establishes that the forests existing in the national territory, taken together, constitute a common interest to all inhabitants of the country. The Code of 34 was in force until 1965 and was revoked by the Forest Code, through Law 4,771/1965. According to Gonzáles and Bacha (2006), the Second Code emerged as a new regulatory policy that became more active in controlling deforestation, seeking to create effective conditions for the development of a reforestation policy.

This code remains in force in the country, having undergone a series of changes over four decades. It is characterized by the definition of permanent preservation areas and legal reserve, which are areas within the rural property, of private property, of limited use. The percentages of legal reserve are defined according to Brazilian biomes ranging from 20% to 80 in areas located in the Legal Amazon. Currently, the Forest Code stands out in the country for being the stage of several discussions in the legislature due to the new amendments proposals.

In Brazil, in the 1970s, several policies for rural development and occupation of the most remote regions were enacted. Because of this and the United Nations Conference on the Environment, which took place in 1972, there is a concern in the country about water resources and forests, influencing the creation of the first Brazilian environmental agency, the Special Environment Secretariat (SEMA), in 1973.

In the 1980s, due to international pressures, these concerns were intensified, resulting in the establishment of the National Policy for the Environment (PNMA), Law 6.938 / 1981 and Article 225 of the 1988 Federal Constitution dedicated to guaranteeing everyone the right to the environment balanced.

Despite the extensive forest area in Brazil, it was only through Decree 3.420 / 2000, that the National Forest Program (PNF) was established with the objective of articulating sectorial public policies to promote sustainable development, reconciling use with conservation of forests. Brazilian forests. This Program is coordinated by MMA and consists of projects carried out in a participatory and integrated manner by the federal, state, district, municipal governments and organized civil society.

With regards to public forests, Law 11.284 / 2006 was enacted, which provides for the management of these areas for sustainable production, establishing the Brazilian Forest Service (SFB) to exercise the function of managing body. As a consultative body, the Public Forest Management Commission was created, within the scope of the Ministry of the Environment. The same law establishes the National Forest Information System (SNIF) which aims to integrate and unify information, to ensure transparency and publicity about forest management in the country, as well as the monitoring of programs and actions developed by public institutions for the management of forests.

Also, by law 11.284, the control and inspection of public forests, at the federal level, is the responsibility of the Brazilian Institute for the Environment and Renewable Natural Resources (IBAMA). Subsequently, with the creation of the Chico Mendes Institute for Biodiversity Conservation (ICMBio), by Law 11,516 of August 28, 2007, this body starts to exercise the role of inspecting and monitoring within the Conservation Units - territorial spaces with the objective of talking to the natural resources, legally instituted by the public.

Thus, the large number of public institutions active in Brazilian forest management is emphasized, with few instances of social participation being established. In this sense, the Federal Community and Family Forest Management Program, established by Decree 6874/2009, which through the Ordinance establishes the creation of a Working Group, composed of representatives of the government and civil society, to contribute for preparing the Annual Plans.

5. Economic Sustainability, Integration, Governance and Policies

In terms of economic sustainability, as sources of funds can be mentioned: primary resources (concrete sources or firm revenues); derived resources (sources and financing mechanisms resulting from sectoral and regional development policies); traditional resources (financing resources); private resources (private banks, capital markets, insurance, futures markets) and international resources (resources from international financing agencies).

Preliminary studies show that the potential for collecting water resources in Brazil is R $ 520 million / year, with a large concentration in the Paraná River Basin. However, the implementation of the collection entails several previous steps, in addition to relatively high transaction costs.

In addition, the following possibilities for expanding the resources and financial sustainability of SINGREH follow: advances in collection, both in terms of methodology, and in the implementation in new basins; improvement in the implementation of the PNMC, especially in mitigation and adaptation actions to climate change and REDD+; improving the articulation of water resources management with forest management; Improvement of the articulation of water resources management with land use and conservation management; greater links with States, Municipalities, state and sector funds and banks.

Regarding the sources of funds for forest development, the following stand out: National Fund for Forest Development (FNDF), Amazon Fund and National Environment Fund (FNMA).

As for the management of water resources in Brazil, this has developed in a fragmented and centralized way. In addition, each user sector (electricity, navigation, irrigated agriculture, sanitation) carried out its planning and actions in an individualized and disconnected manner. Since the 1980s, there has been an increasing progress towards decentralization and the participation of society in the decision-making process. In this sense, numerous technical tools of management (instrumental rationality) and negotiation (dialogical rationality) have been used in the instruments of Law 9.433 / 1997. Such instruments should be the stage for constant improvement and incorporation of new scientific and technological advances.

In the case of forestry policy, in contrast to the water resources policy, it has been fragmented into several public bodies and has been decentralized since the last decade.

Finally, it appears that there must be greater integration between the PNRH and the PNMC and the Forest Policy. Therefore, an increase in the participation of civil society should be sought, in quantity and quality, in the Basin Committees, in the State Councils, in the National Water Resources Council - CNRH, in the Brazilian Forum on Climate Change and in the Management Commission Public Forests.

6. References

BOFF, L.: Ética e gestão das águas. Palestra proferida no Seminário "Água, Desenvolvimento e Justiça Ambiental", Brasília, MMA, 2003.

FREITAS, M. A. S.: A Regulação dos Recursos Hídricos. 1. ed. Rio de Janeiro: CBJE, 2009. 174 p.

FREITAS, M. A. S.: Que venha a seca: modelos para a gestão de recursos hídricos em regiões semiáridas, Rio de Janeiro: Ed. CBJE, 417p, 2010.

GONZÁLES, M. V. e BACHA, C. J. C.: Um Estudo Comparativo entre as Políticas Florestais do Brasil e Paraguai. XLIV CONGRESSO DA SOBER- Questões Agrárias, Educação no Campo e Desenvolvimento. Fortaleza, 23 a 27 de Julho de 2006.

HADDAD, P.R.: 2050 – petróleo ou água? O Estado de São Paulo, em 13/05/2008.

LANNA, A.E.: Plano Nacional de Capacitação em Recursos Hídricos, Curso "Introdução à Gestão de Recursos Hídricos", MMARHAL/SRH, Vol. II, Tomo I, 1997.

LOBATO DA COSTA, F. J.: Estratégias para o gerenciamento dos recursos hídricos no Brasil: Áreas de cooperação com o Banco Mundial. Brasília-DF: BIRD, abr. de 2003.

MMA: Água: Manual de Uso – Vamos cuidar de nossas águas – Implementando o Plano Nacional de Recursos Hídricos, 2ª edição, Brasília – DF, 2008.

PORTO, M. & LOBATO DA COSTA, F.J.: Mecanismos Econômicos, Sociais e Ambientais de Gestão da Água, Revista Rega, vol. 1, n.2, jul.-dez. 2004

CHAPTER 15

Forest Bioeconomy as an Engine for Sustainable Development and Mitigation of the Effects of Climate Change

Sandra Regina Afonso

PhD in Forestry Sciences - UnB; Researcher at the Brazilian Forest Service - SFB; sandra.afonso@florestal.gov.br

Abstract: Brazil presents a great opportunity for the development of the bioeconomy, starting from the management of natural forests, especially public forests, as well as from the integration of the forest component to agricultural systems, especially in private areas. With regards to the management of natural forests, the importance of expanding the use of biodiversity products, especially non-timber products in a sustainable manner and with technological innovation, is highlighted. Currently, only 10 products account for more than 90% of non-timber forest production from native forests. A potential that is still underutilized, especially if we consider the Amazon biome. Regarding the integration of the forestry component into agricultural systems, several forms of production that are being developed all over the world stand out, highlighted here in this chapter, opportunities that are even more interesting when it comes to the Cerrado biome. Finally, these development opportunities from the Forest Bioeconomy stand out as paths to Sustainable Development and the Mitigation of the Effects of Climate Change.

Keywords: Forest Bioeconomy, management of natural forests, non-timber products, sustainable development

1. Introduction

Forest Bioeconomy was defined by Afonso e Freitas (2021, in press) as a set of economic activities based on products and forest ecosystem services, carried out in a sustainable and innovative way, and taking into account also environmental, cultural and social aspects associated to the forest resources.

This definition is adapted to the Brazilian reality, considering that Brazil is a country with 58% of the territory occupied by forests. It has the second largest area of forests in the world, corresponding to about 12% of the total forests on the planet (FAO, 2020). Forests that are home to an unparalleled biodiversity associated with a set of social, economic and cultural aspects. This characteristic points to a great potential for the development of the forest bioeconomy in Brazil.

According to data from the National Forestry Information System (SNIF), in 2019, Brazil had an area covered by forests estimated at 484.1 million hectares, with 474.1 million hectares covered by natural forests (98%) and 9, 9 million hectares (2%) per planted (SNIF, 2021).

With regard to planted forests, the value of production, accounted by IBGE, in 2019, was about U$ 3 billion, 77.7% of the value of forest production, referring to the commercialization of pine and eucalyptus and, to a lesser extent scale, other forestry species.

With regard to the extraction of natural forests in Brazil, the country with the greatest diversity on the planet, the value was around U$ 1 billion, with around 65% of this value referring to wood products and 35% related to non-wood forest products.

It is observed that the management of natural forests still has a great potential for increasing and promoting development, since they add up to 474.1 million hectares, which means that 55% of the Brazilian territory is covered by natural forests. This amount has the potential to offer ecosystem services, timbers and non-timber forest products, and thus contribute to food security and income generation for the communities living in these areas, as well as for the mitigating the effects of climate change.

In addition to the management of existing forests, new forms of production, including the forestry component in production systems, have been identified worldwide. In this chapter we will discuss the use of forests based on their sustainable management, as well as in an integrated way with agricultural systems.

2. Forests as Drivers for Sustainable Development and Mitigation of the Effects of Climate Change

In Brazil, the forest production of natural forests comes from public and private forest lands. The National Register of Public Forests (CNFP), comprises all federal states, and municipal public forested lands. It includes areas designated to Indigenous Peoples, conservation units, and other public forests located in urban or rural areas. The distribution per biome of federal and state public forests included in the National Register of Public Forests - CNFP, in 2018, can be shown in Table 1.

Table 1: Federal and State Public Forests in Brazil

Biome	Area (million ha)	Public Forests total area %
Amazon	284,98	92,2
Caatinga	1,62	0,5
Cerrado	17,35	5,6
Atlantic Forest	4,03	1,3
Pampa	0,15	0
Pantanal	1,06	0,3
Total	**309,2**	**100**

Public forests account for 309.2 million hectares and are almost entirely present in the Amazon biome. Of the total public forest areas, about 50% are for community use and have great social and economic relevance since they generate wood and non-wood forest products. With regards to non-timber forest production, it has been increasingly evidenced, not only by community use, but also by its commercial value.

In Brazil, according to IBGE data, seven non-timber forest products stand out for their economic importance and which corresponds to more than 90% of the total production value, namely: açaí fruit, yerba mate, carnauba powder, Brazil nut, babassu almond, Araucaria seed, pequi fruit, Palm Heart, carnauba wax and umbu fruit. This production comes from public and private forests.

With regards to private forests, the areas of Legal Reserve stand out, which, according to the new Forest Code (Law 12.651 / 2012) are those areas located within rural properties, where natural vegetation must be maintained. These areas, mostly forests, have the function of promoting the conservation of biodiversity and can be used through sustainable forest management to produce goods and services. National data point to the existence of approximately 120 million hectares of Legal Reserves registered in approximately 6 million rural properties.

Rodrigues et al. (2009) highlight the importance of biodiversity conservation in the remnants of private properties that can present themselves as holders of biodiversity, if they are adequately protected and recovered, with actions, among others, for the management and enrichment of species.

Law 12.651 of May 25, 2012 differentiates the areas occupied by family farmers or traditional peoples and communities, allowing them to practice agroforestry in areas of Legal Reserve, as long as it does not de-characterize the existing vegetation cover and does not harm the environmental function of the area . In view of this, agroforestry systems are presented as an alternative for their potential to generate income and promote various environmental services (Miccolis et al., 2016).

According to the World Agroforestry Center (ICRAF), agroforestry systems (SAFs) are systems based on dynamics, ecology and management of natural resources that, through the integration of trees on the property and the agricultural landscape, diversify and sustain production with greater social, economic and environmental benefits for all those who use the soil at different scales (Miccolis et al., 2016).

There are no fixed models for the establishment of agroforestry systems, however, there are guidelines for building adaptable solutions combining technical and empirical knowledge (Miccolis et al., 2017). Farmers often have extensive knowledge of propagation methods and suitability for specific light and soil conditions for a variety of crop and tree species that contribute to restoration projects (VIEIRA et al., 2009).

In this context, a strategy envisaged for the conservation of biomes is the expansion of areas covered with native species of economic value in territories occupied by family farmers or by traditional peoples and communities, through the implantation or enrichment of agroforestry systems. Thus, encouraging the agroextractive practice, it is expected to bring economic benefits to producers and ecologists in a broader way for all biomes.

According to Vieira et al. (2009) agroforestry systems can be used as a transition phase in forest restoration helping to connect farmers with restoration practice. The planting of annual crops combined with tree species contributes to the survival and growth of both types of species.

In this context, a series of concepts have been discussed and have been subsidizing practices around the world, such as: "domestic forest", "forest gardens", "climate smart agriculture" and "integrated landscape management". The term "domestic forest" highlights the close relationship that the domestication process establishes between a specific human group and its forest areas - which are managed to meet the diverse needs of that group. In these areas, various forest management practices and cultivation of forest species are developed in an integrated manner with agriculture - thus creating spaces with particularly characteristics.

Michon et al. (2007) analyzed several studies carried out by authors from Southeast Asia and Africa and found that the integration between forest management and agriculture was the reason for the development of "domestic forest". These spaces provide means of subsistence, as well as being related to the culture and socio-political relations of managers. In this way, it integrates production and conservation with social, political and spiritual dimensions.

This concept is similar to the concept of "forest gardens", which are complex agroforestry systems, with different strata, characterized by high diversity, including perennials at all levels, from tall trees to short trees, shrubs, herbs, soil covers, tubers and creepers. Björklund et al. (2018) studied 12 experiences in Sweden and concluded that these spaces provide fresh products for consumption throughout the year, as well as becoming beautiful environments for interaction and learning. Similar practices are reported in the Cerrado of Minas Gerais, like the productive yards established in the lots of the American Agroextrivist Settlement, as described in the work by Carvalho and Bergamasco (2016).

The concept of "climate smart agriculture" brings another approach that has recently achieved great prominence, due to the challenges of adapting and mitigating climate change. According to Scherr et al. (2012), "climate smart agriculture" emerged bringing the message that agricultural systems can be developed and implemented to simultaneously: guarantee food security and rural livelihoods; facilitate adaptation to climate change; and provide mitigation benefits from these changes. The development of this concept was conducted by international institutions, particularly the United Nations Food and Agriculture Organization (FAO) and the World Bank. The Consultative Group on International Agricultural Research (CGIAR) led this discussion and the concept has now been incorporated into projects financed by the World Bank.

Climate smart agriculture has three objectives: to increase productivity to improve food security and rural development; decrease greenhouse gas emissions and increase carbon sinks; and expanding the capacity to act at various levels - from local to global (Campbell, et al. 2014). In this way, FAO takes an ecosystem approach, working on a landscape scale and encouraging intersectoral cooperation. The World Bank, in turn, includes the concept of "integrated landscape management" (in Portuguese, integrated landscape management) as a strategy for political action in favor of agricultural development and ecosystem conservation (Scherr et al., 2012).

Integrated landscape management approaches work deliberately to support food production, ecosystem conservation and rural livelihoods across entire landscapes. The ways of acting are already known under several terms, such as: ecoagriculture, landscape restoration, territorial development, model forests, integrated management of river basins, agroforestry systems and the ecosystemic approach to the management of agricultural systems, among many others (Scherr et al., 2012).

Considering the forms of action addressed and in view of the various concepts discussed worldwide, it is highlighted that, in addition to the importance of the agroforestry systems previously presented, it is necessary to speak about silvopastoral systems in the Cerrado biome. As defined by Porfírio-da-Silva (2004) "silvopastoral system is the intentional combination of trees, pasture and cattle in the same area at the same time and managed in an integrated manner, with the aim of increasing productivity per unit area". According to the same author, silvopastoral systems have economic and environmental benefits for producers and society. They are multifunctional systems, where there is the possibility of intensifying production through the integrated management of natural resources, avoiding their degradation, in addition to recovering their productive capacity.

The silvopastoral system exists in several countries with forest and savanna ecosystems and was formed by selective deforestation, conduction of natural regeneration and, less frequently, by plantations (BRUZIGUESSI, 2016; PYWELL, 2015; SHANLEY, 2005). In Brazil, the inclusion of native tree species in silvopastoral systems is still poorly studied.

3. Final Considerations

In general, it is observed that Brazil presents a great opportunity for the development of the bioeconomy, starting from the management of natural forests, especially public forests, as well as from the integration of the forest component to agricultural systems, especially in private areas.

With regards to the management of natural forests, the importance of expanding the use of biodiversity products, especially non-timber products in a sustainable manner and with technological innovation, is highlighted. Currently, only 10 products account for more than 90% of non-timber forest production from native forests. A potential that is still underutilized, especially if we consider the Amazon biome.

Regarding the integration of the forestry component into agricultural systems, several forms of production that are being developed all over the world stand out, highlighted here in this chapter, opportunities that are even more interesting when it comes to the Cerrado biome.

Finally, these development opportunities from the Forest Bioeconomy stand out as paths to Sustainable Development and the Mitigation of the Effects of Climate Change.

4. Bibliography

AFONSO S. R., FREITAS, J. V. The Brazilian Non-Timber Forest Production and the transition for the Bioeconomy. In: The Bioeconomy and Non-Timber Forest Products: Theory and Empirical Advances. 2021

BJÖRKLUND, J.; EKSVÄRD, K.; SCHAFFER, C. Exploring the potential of edible forest gardens: experiences from a participatory action research project. Agroforest Syst, 2018. URL: https://doi.org/10.1007/s10457-018-0208-8

BRUZIGUESSI, E. P. (2016). Árvores nativas do cerrado na pastagem: Por quê? Como? Quais? Tese de doutorado. Publicação PPGEFL. TD - 299/ 2015, Programa de Pós-Graduação em Ciências Florestais, Universidade de Brasília - UnB, Brasília, DF, 163p.

CAMPBELL, B. M.; THORNTON, P., ZOUGMORÉ, R., ASTEN, P.; LIPPER, L. Sustainable intensification: What is its role in climate smart agriculture? Current Opinion in Environmental Sustainability 2014, 8:39–43. 2014. URL: https://www.sciencedirect.com/science/article/pii/S1877343514000359

CARVALHO, I. S. H. M; BERGAMASCO, S. M. M. P. P. Assentamento Agroextrativista Americana: Campesinato, Biodiversidade e Agroecologia no Cerrado Mineiro. Retratos de Assentamentos v.19, n.1, 2016

FAO (Food and Agriculture Organization of the United Nations) (2020) Global Forest Resources Assessment 2020: Main report. 1-186. Rome: FAO.

LAW 12.651 / 2012. Lei 12.651, de 25 de maio de 2012 que dispõe sobre a proteção da vegetação nativa. Available in:

<hhttp://www.planalto.gov.br/ccivil_03/_ato20112014/2012/lei/l12651.htm > Acessed 10 Jan 2021.

MICCOLIS, A.; PENEIREIRO, F. M.; MARQUES, H. R.; VIEIRA, D. L. M.; ARCO-VERDE; M. F.; HOFFMANN; M. R.; REHDER, T.; PEREIRA, A. V. B. Restauração Ecológica com Sistemas Agroflorestais: como conciliar conservação com produção. Opções para Cerrado e Caatinga. Brasília: Instituto Sociedade, População e Natureza – ISPN/Centro Internacional de Pesquisa Agroflorestal – ICRAF, 2016.

MICCOLIS, A.; PENEIREIRO, F. M.; VIEIRA, D. L. M.; MARQUES, H. R.; HOFFMANN; M. R. Restoration through agroforestry: options for reconciling livehoods with conservation in the Cerrado and Caatinga Biomes in Brazil. Experimental Agriculture, 1-18, 2017

MICHON, G., H. DE FORESTA, P. LEVANG, AND F. VERDEAUX. Domestic forests: a new paradigm for integrating local communities' forestry into tropical forest science. Ecology and Society 12(2): 1. 2007. [online] URL: http://www.ecologyandsociety.org/vol12/iss2/art1/

MIMENZA, H. E. Dispersed trees in pasturelands of cattle farms in a tropical dry ecosystem. Tropical and subtropical agroecosystems, v. 14 p. 933-941, 2011.

PORFÍRIO-DA-SILVA, V. Sistemas Silvipastoris. Embrapa Florestas. 2004. URL: https://www.cnpf.embrapa.br/pesquisa/safs/

PYWELL R.F., HEARD M.S., WOODCOCK B.A., HINSLEY S., RIDDING L., NOWAKOWSKI M., BULLOCK J.M. 2015 Wildlifefriendly farming increases crop yield: evidence for ecological intensification. Proc. R. Soc. B 282: 20151740. http://dx.doi.org/10.1098/rspb.2015.1740

RODRIGUES, R. R.; BRANCALION, P. H. S.; ISERNHAGEN, I. Pacto pela restauração da mata atlântica: referencial dos conceitos e ações de restauração florestal [organização edição de texto: São Paulo: LERF/ESALQ: Instituto BioAtlântica, 2009.

SCHERR, S. J.; SHAMES, S.; FRIEDMAN, R. From climate-smart agriculture to climate-smart landscapes. Agriculture & Food Security 1:12. 2012. URL: https://doi.org/10.1186/2048-7010-1-12

SHANLEY, P.; PIERCE A.; LAIRD, S. Além da Madeira: certificação de produtos florestais não-madeireiros. Bogor, Indonésia: Centro de Pesquisa Florestal Internacional (CIFOR), 2005.

SNIF (Sistema Nacional de Informações Florestais) (2020). Serviço Florestal Brasileiro. 2019. Available in: < h http://snif.florestal.gov.br/pt-br/> Acessed 10 Jan 2021.

VIEIRA, D. L. M.; HOLL, K.D.; PENEIREIRO, F. M. Agro-Successional Restoration as a Strategy to Facilitate Tropical Forest Recovery. Restoration Ecology Vol. 17, No. 4, pp. 451–459, 2009

CHAPTER 16

Overturn of Integumentary Dormancy of Seeds for the Forest Restoration in Brazilian Savanna

Marcos Airton de Sousa Freitas[1] & Sandra Regina Afonso[2]
[1] Specialist in Water Resources at the National Water Agency - ANA; Ministry of Regional Development - MDR; masfreitas@ana.gov.br
[2] PhD in Forestry Sciences - UnB; Researcher at the Brazilian Forest Service - SFB; sandra.afonso@florestal.gov.br

Abstract: The Brazilian savanna is among the biomes with the highest floristic diversity on the planet. However, 48% of the native area was removed or altered due to land occupation for agriculture and urbanization. Therefore, there is demand for recovery of the biome. The jatobá (Hymenaea courbaril), the mutamba (Guazuma ulmifolia) and the canzileiro (Platypodium elegans) are species used in the restoration, through the planting of seedlings. However, the seeds of these species show integument dormancy necessitating the application of treatments to break dormancy. Different treatments were evaluated aiming at the acceleration, the uniformization of the germination and the increase of the germination rates. Pre-germination treatments were used: mechanical scarification with subsequent soaking in water for 48h and imbibition in hot water (100ºC). For each species, the Emergency Speed Index (ESI) and the percentage of accumulated germination were calculated, in both cases, the logistic model was successfully adjusted. Furthermore, the Kaplan-Meier survival analysis was employed. The seeds submitted to treatment obtained higher percentage of germination, higher ESI and higher levels of survival, when compared to the control. The relation between the accumulated precipitation and the germination rate was also analyzed, showing that there was a good correlation between these variables.

Keywords: Emergence; Plant propagation; Savanna; Seedling production.

1. Introdução

The Brazilian savannah is among the biomes with the greatest floristic diversity on the planet. However, 48% of the native area has been removed or altered due to land occupation for agriculture and urbanization. Therefore, there is a demand for recovery of the biome.

To restore the resilience of degraded environments, several Programs for the Recovery of Degraded Areas (PRAD) and reforestation have been developed. Thus, several techniques, among them, the planting of seedlings and sowing of native species, were created to accelerate the recovery and restoration process.

Jatobá (*Hymenaea courbaril*), mutamba (*Guazuma ulmifolia*) and canzileiro (*Platypodium elegans*) are some of the species used in the restoration, through the planting of seedlings. The seeds of these species, however, have integumentary dormancy requiring the application of treatments to break dormancy.

In this sense, seed germination and seedling survival are the most critical steps in a natural plant community (Kitajima; Fenner, 2000). They are fundamental in the restoration of ecosystems, which depends, above all, on the reintroduction of species in the environment to be restored. On the other hand, the knowledge about the germination of Cerrado (Brazilian savanna) seeds is still incipient in view of its enormous floristic wealth, which presents itself as one of the first obstacles to restoration, especially in view of the accelerated pace of conversion of its lands for agricultural or pastures (Klink; Machado, 2005).

For seedling producers of native species and nurseries, the dormancy mechanism has a disadvantage, since it induces a non-uniformity between the seedlings and a greater demand for time in production, in addition to a greater risk of loss of seeds due to deterioration, as they remain more time in the soil before germination (Eira et al., 1993).

2. Material and methods

This chapter analyzes and discusses the process of overcoming integumentary dormancy of these three species used in restoration, through the planting of seedlings, aiming at forest restoration in the cerrado: the jatobá (*Hymenaea courbaril*), the mutamba (*Guazuma ulmifolia*) and the canzileiro (*Platypodium elegans*).

The jatobá (*Hymenaea courbaril*) is a very showy tree species, found throughout America, belonging to the family Leguminosae (*Fabaceae*), subfamily *Caesalpinoideae*. In Brazil, it extends from Piauí to the north of Paraná. According to Moreira et al. (2005), and also Cabral et al. (2015), jatobá beyond ecological importance, since it participates in the composition of heterogeneous reforestation and the afforestation of parks and large gardens (Lorenzi, 1992), and has an agronomic potential for the use of the stem and fruits (Cruz; Carvalho; Oliveira, 1997). Being a supplier of wood for sawmills, construction and poles; it supplies phytochemical and medicinal products, besides serving for shading, hedges, bee flora, human and animal food. (Caramori et al., 2004).

According to Melo and Polo (2007), jatobá is a species with great forest and environmental importance because it is potentially fixing and storing carbon, and for its scenic beauty and economic importance. Melo and Polo (2007) concluded in their work that the jatobá seeds have an impermeable wrap that causes certain dormancy, and the scarification of the seeds is necessary to break this dormancy.

The seeds, according to Cardoso (2004), tend to have some dormancy mechanism. In spite of being considered beneficial with regard to the process of survival of plant species, for the nurseryman or producer of seedlings, dormancy is a harmful process to the production of seedlings and needs to be overcome in order to obtain uniform and rapid emergence. In view of the need for native vegetation replacement or recovery of deforested areas, rational forest restoration has become fundamentally important. Among the various factors to be studied, there is one particularly that directly affects the production of seedlings, which the seed dormancy process. With the objective of this work to evaluate the effect of immersing Jatobá seeds in hot water at a temperature of 50º C at different times of exposure (Moreira et al., 2005).

Lima et. al. (2014) aiming to investigate the influence of different light conditions on the germination of fifteen species of the Cerrado in a nursery condition, showed that light is a factor that influences differently the species of the Cerrado, which may or may not produce effects on germination. Ratifying other previous studies (Zaidan; Carreira, 2008; Rossato; Kolb 2010; Mota et al., 2012), ten among the fifteen species studied did not show different germination due to the availability of light.

3. Results and discussion

The fruits were harvested from matrices located in the cerrado region of Condomínio Verde and Fazenda Taboquinha, belonging to the Ribeirão Taboca watershed (Figure 1), in Brasília, Distrito Federal. The fruits were processed and the seeds extracted and planted.

Figure 1: Ribeirão Taboca hydrographic basin

When germinating each species, the germination percentages and the Emergence Speed Index (ESI) (Maguire, 1962) for each treatment were calculated. The ESI is given by the formula:

$$IVE = \frac{G_1}{N_1} + \frac{G_2}{N_2} + \dots + \frac{G_n}{N_n}$$

where

G1, G2, Gn = number of seedlings in the first, second and last count
N1, N2, Nn = number of sowing days at the first, second and last count.

The germination speed was calculated by counting the number of seeds germinated per day, between the fifth and the 30th day of installation of the test. The formula used to calculate the Emergence Speed Index (ESI) is shown below:

$$IVG = \sum_{i=1}^{30} \frac{G_i}{i}$$

G = number of seeds germinated i days after installation of the test
i = Number of days after installing the test

The germination capacity was calculated in percentage of seeds germinated after the installation of the test, in relation to the total of seeds sampled. The seeds were considered to have germinated from the moment the radicle length exceeded the seed length. Figure 2 shows the Emergency Speed Index (ESI) for the jatobá (*Hymenaea courbaril*), both with scarification and for cases without scarification. It appears that, with scarification, the ESI reached values of almost 80%. A logistic curve was adjusted for the case with scarification, with equation $y = a / (1 + b * \exp(-cx))$. The coefficient values are, as follow: a = 0.77603783; b = 12473681 and c = 0.59824196. The correlation coefficient was r = 0.99421920. A logistic curve was fitted for the case without scarification, with equation $y = a / (1 + b * \exp(-cx))$. The coefficients values, in this case, are as follow: a = 0.50114031; b = 3712069.3 and c = 1.3245462. The correlation coefficient r = 0.99907110.

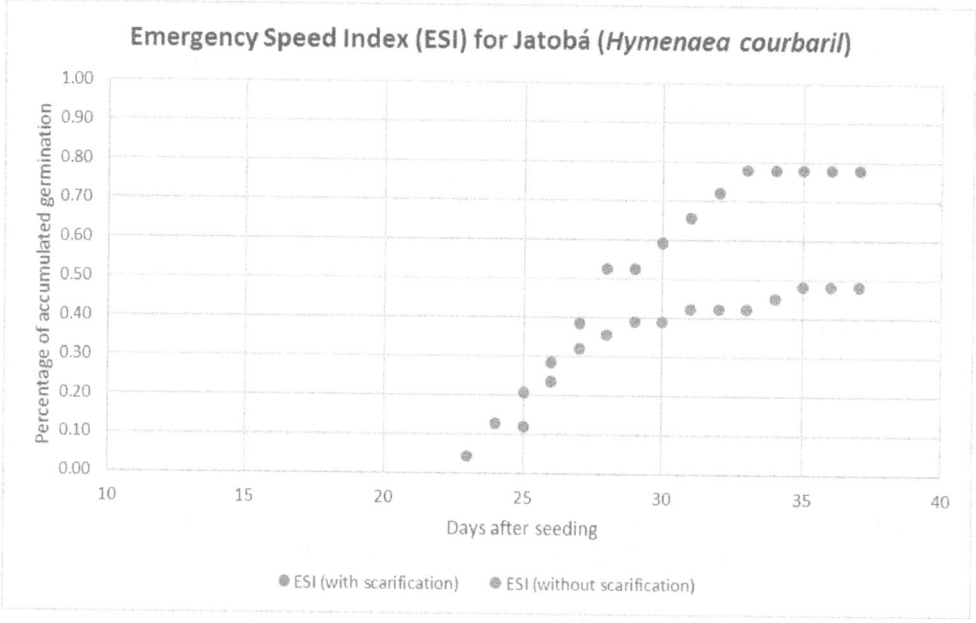

Figure 2: Emergency Speed Index (ESI) of the jatobá (*Hymenaea courbaril*).

The water content has a pronounced influence on the physical and chemical properties of forest seeds, and this determination is very important in all stages of the seed technology process, from handling, processing, storage, among others (Carvalho, 2005).

Pilon et al. (2012), analyzing the germination of *Tachigali vulgaris* (carvoeiro) seeds, also belonging to the Leguminosae (*Fabaceae*) family, subfamily *Caesalpinoideae*, describe that both chemical scarification with sulfuric acid and mechanical scarification with sandpaper resulted in a higher germination rate at 70% in seven days, which supports the unequivocal recommendation that *T. vulgaris* seeds should be scarified before sowing, as recommended by Carvalho (2005) and Souchie (2011).

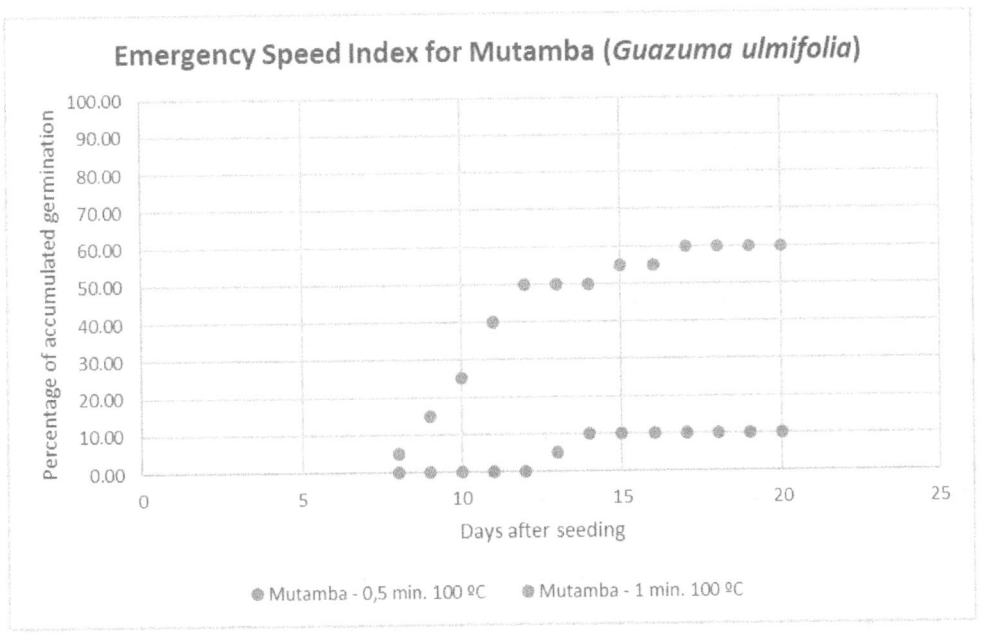

Figure 3: Emergency Speed Index (ESI) of the mutamba (*Guazuma ulmifolia*).

Figure 2 shows the Emergency Speed Index (ESI) for the mutamba (*Guazuma ulmifolia*), when a thermal shock was used in the seeds, both for the case of 0.5 minutes in water with a temperature of 100ºC, and for the case of 1, 0 minutes in water with a temperature of 100ºC. It appears that, with the thermal shock of 1.0 minutes in water with a temperature of 100ºC, the ESI reached values of the order almost 60%, while for the thermal shock of 0.5 minutes in water with a temperature of 100ºC, the ESI reached values of only 10%. A logistic curve, with equation $y = a / (1 + b * \exp(-cx))$, was adjusted for the one minute thermal shock in water with a temperature of 100ºC. The coefficients are as follow: $a = 90.316816$; $b = 49048128$ and $c = 0.64191002$. The correlation coefficient was $r = 0.99023354$. For the case 0.5 minutes in water with a temperature of 100ºC was adjusted, also a logistic curve, with equation $y = a / (1 + b * \exp(-cx))$. The coefficients are, in this case, as follow: $a = 50.157112$; $b = 2595000.1$ and $c = 1.2812946$. The correlation coefficient $r = 0.99867525$.

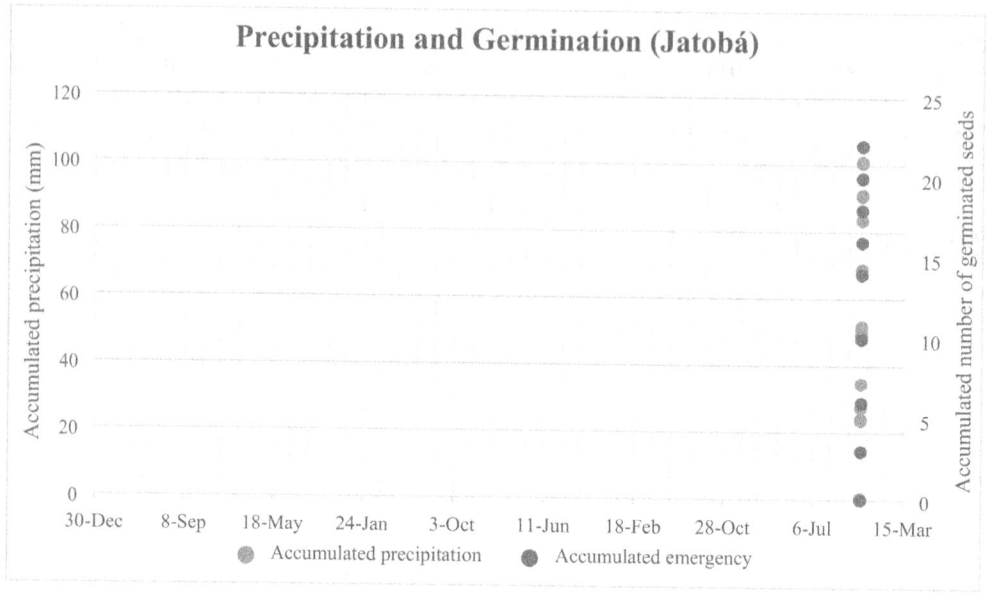

Figure 3: Accumulated Precipitation and Germination for Jatobá.

Figure 3 shows the relationship between accumulated precipitation (in millimeters) and the accumulated number of germinated seeds, for jatobá, showing that there was a good correlation between these variables.

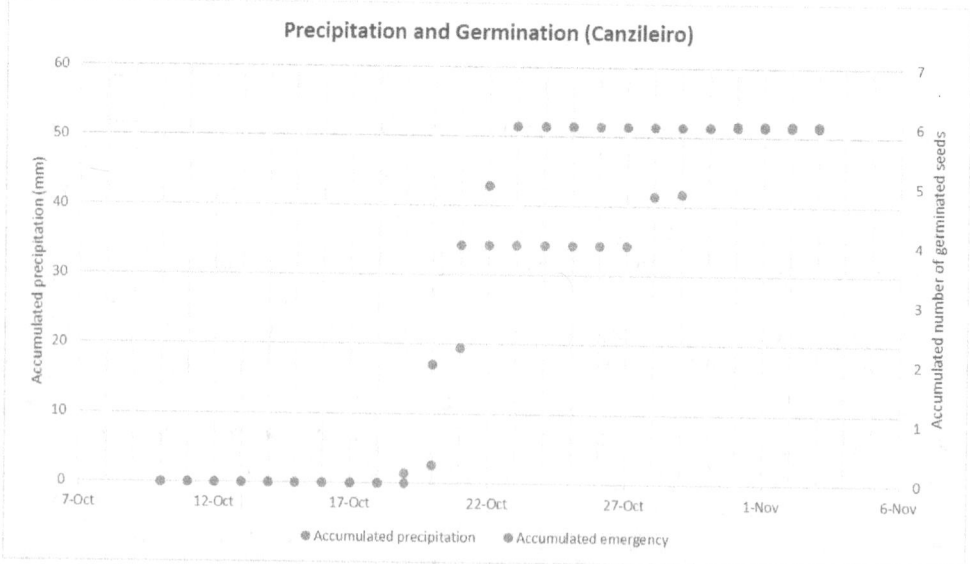

Figure 4: Accumulated Precipitation and Germination for the Canzileiro.

Figure 4 shows the relationship between the accumulated precipitation (in millimeters) and the accumulated number of germinated seeds, for the caneleiro, showing that there was a good correlation between these variables.

The statistical techniques known as survival analysis are used when it is intended to analyze a phenomenon in relation to a period of time, that is, the time elapsed between an initial event, in which a subject or an object enters a particular state and an event final, which changes this state (Bustamante-Teixeira et al., 2002).

In the analysis of survival by the Kaplan-Meier method (Kaplan & Meier, 1958; Lee, 1992; Kleinbaum, 1995; Carvalho et al., 2011) the time intervals are not fixed, but determined by the appearance of a failure (for example, death).

In this situation, the number of deaths in each interval must be one. This is a non-parametric method, that is, it does not depend on the probability distribution (Colton, 1979), and to calculate the estimators, first, you must order the survival times in ascending order ($t1 \leq t2 \leq ... \leq tn$).

The function S (t) is defined by an estimator known as the Kaplan-Meier limit product estimator, as it is the limit of the product of terms up to time t:

$$S(t) = \prod_{t=0}^{j} \frac{l_j - i}{l_j}, wo \{i = 1, for\ failure; i = 0, if\ not failure$$

and lj = number of plants exposed to risk at the beginning of the period.

Thus, in Figure 5, we find the Kaplan-Meier curve calculated for the jatobá, for cases with and without scarification.

In Figure 6, it is possible to verify the seedlings after planting in the region belonging to the Ribeirão Taboca hydrographic basin, right after a period of burning. The seedlings were planted in Condomínio Verde, in springs areas, aiming at forest recovery, as well as to supply the carbon neutralization associated with the Mexidão Cultural Project, in the last three years (2017 to 2019).

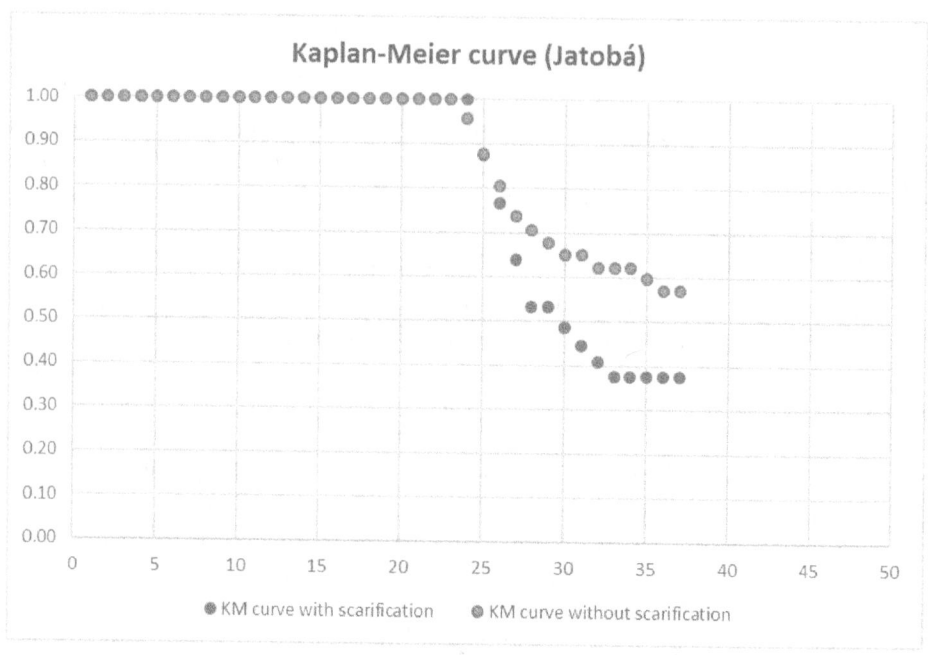

Figure 5: Kaplan-Meier curve for Jatobá.

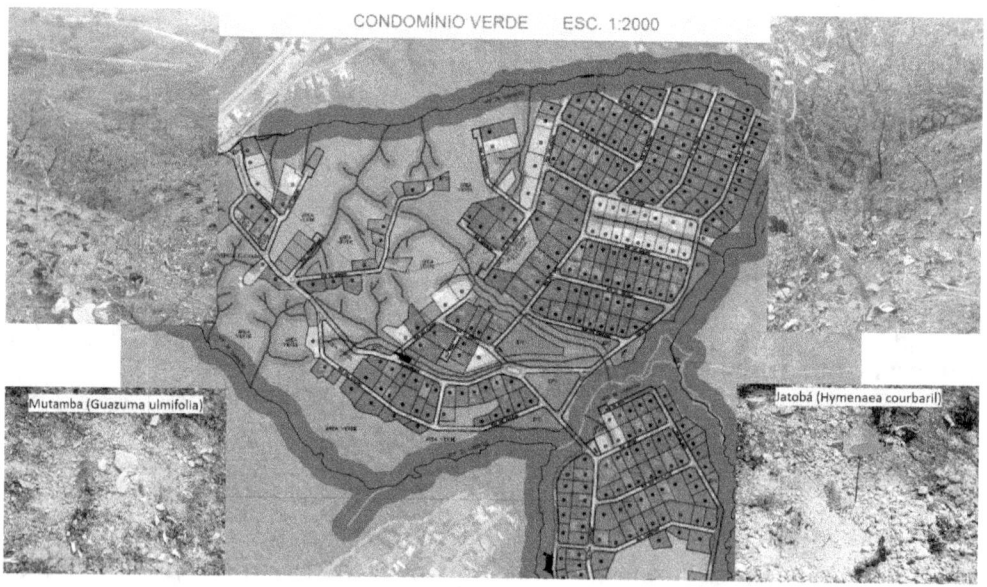

Figure 6: Burnt and reforested area in the Ribeirão Taboca hydrographic basin

4. Conclusions

Different treatments were evaluated aiming at acceleration, uniformity of germination and increased germination rates. Pre-germinative treatments were used: mechanical scarification with subsequent soaking in water for 48 hours and soaking in hot water (100ºC). For each species, the Emergency Speed Index (ESI) and the percentage of accumulated germination were calculated, adjusting, in both cases, the logistic model, with success. In addition, Kaplan-Meier survival analysis was used. The seeds submitted to treatment obtained a higher percentage of germination, higher ESI and higher levels of survival, when compared to the control. The relationship between accumulated precipitation and germination rate was also analyzed, showing that there was a good correlation between these variables.

5. References

BUSTAMANTE-TEIXEIRA, M.T.; FAERSTEIN, E.; LATORRE, M.R. Técnicas de análise de sobrevida – Survival analysis techniques. Cad. Saúde Pública, Rio de Janeiro, 18(3):579-594, mai-jun, 2002.

CABRAL, E. M. S.; CASTILHO, R.M.; PAGLIARINI, M. K. Germinação de sementes e desenvolvimento de mudas de Jatobá (Hymeneae courbaril L. var. Stilbocarpa), Revista Eletrônica Thesis, São Paulo, ano XI, n. 23, p.16-28, 1° semestre, 2015.

CARAMORI, S. S.; LIMA, C. S.; FERNANDES, K. F. Biochemical characterization of selected plant species from Brazilian savannas. Brazilian Archives of Biology and Technology, v. 47, n. 2, p. 35-42, 2004.

CARVALHO M.S. et al. 2011. Análise de sobrevivência: teoria e aplicações em saúde. Rio de Janeiro, FIOCRUZ, 2.ed. 432p.

CRUZ, E. D.; CARVALHO, J. E. U.; OLIVEIRA, R. P. Variabilidade na germinação e dormência em sementes de Centrosema pubescen Benth. Pasturas Tropicales, Cali, v. 19, n. 4, p. 37-41, 1997.

EIRA, M. T. S.; FREITAS, R. W. A.; MELLO, C. M. C. Superação de dormência de sementes de Enterolobium contortisilliquuum (VELL.). Morong.–Leguminosae. Revista Brasileira de Sementes, Londrina, v.15, n.2, p.177-82, 1993.

KAPLAN, E. L. & MEIER, P., 1958. Non parametric estimation from incomplete observation. Journal of the American Statistics Association, 53:457-481.

KLEINBAUM, D. G., 1995. Survival Analysis: A Self-Learning Text. New York: Springer.

KLINK, C. A.; MACHADO, R. Conservation of the Brazilian Cerrado. Conservation Biology, Boston, v. 19, n. 3, p. 707-713, jun. 2005.

LEE, E. T., 1992. Statistical Methods for Survival Data Analysis. 2nd Ed. New York: John Wiley & Sons.

MAGUIRE, J. D. Speed of germination-aid in selection and evaluation for seedling emergence and vigor. Crop Science, Madison, v. 2, n. 1, jan./feb. 1962. 176-177p.

MELO, N. C.; POLO, M. Sobrevivência e Germinação de sementes de Hymenaea courbaril L. In: Congresso de Ecologia do Brasil, 2007, Caxambu. Resumos... Caxambu, 2007.

MOREIRA, M.A.T; PAIVA SOBRINHO, S.; SILVA, SJ; SIQUEIRA, A.G. Superação da dormência em sementes de Jatobá. Universidade Estadual de Goiás, Goiás, 6p. 2005.

MOTA, L. H. S.; SCALON, S. P. Q.; HEINZ, R. Sombreamento na emergência de plântulas e no crescimento inicial de Dipteryx alata Vog. Ciência Florestal, Santa Maria, v. 22, n. 3, p. 423-431, jul.-set. 2012.

PILON, N.A.L.; MELO, A.C.G.; DURIGAN, G. COMPARAÇÃO DE MÉTODOS PARA QUEBRA DE DORMÊNCIA DAS SEMENTES DE CARVOEIRO – Tachigali vulgaris L.F. Gomes da Silva e H.C. Lima (FAMÍLIA: FABACEAE – CAESALPINIOIDEAE) (NOTA CIENTÍFICA), Rev. Inst. Flor. v. 24 n. 1 p. 133-138 jun. 2012.

ROSSATTO, D. R.; KOLB, R. M. Germinação de Pyrostegia venusta (Bignoniaceae), viabilidade de sementes e desenvolvimento pós-seminal. Revista Brasileira de Botânica, São Paulo, v. 33, n. 1, p. 51-60, jan.-mar. 2010.

SILVA, L.V.D. et al. (2015). QUEBRA DE DORMÊNCIA DE SEMENTES DE JATOBÁ (Hymenaea courbaril) SOB IMERÇÃO EM ÁGUA QUENTE. IV Congresso Estadual de Iniciação Científica do IF Goiano, 21 a 24 de setembro de 2015.

SOUCHIE, F.F. et al. carvão pirogênico como condicionante para substrato de mudas de Tachigali vulgaris L.G. Silva e H.C. Lima. Ciência Florestal, v. 21, n. 4, p. 811-821, 2011.

ZAIDAN, L. B. P.; CARREIRA, R. C. Seed germination in Cerrado species. Brazilian Journal of Plant Physiology, Londrina, v. 20, n. 3, p. 167-181, jul.-set. 2008.

CHAPTER 17

The Economic-Financial Sustainability of the National Water Resources Management System - SINGREH

Marcos Airton de Sousa Freitas – National Water Agency - ANA,
masfreitas@ana.gov.br

Abstract: The National Water Resources Management System - SINGREH provided for by Item XIX, of Article 21, of the Federal Constitution of 1988, was created by Law No. 9,433, of January 9, 1997. Among its main objectives, the following can be mentioned: coordinate management integrated water; administratively arbitrate conflicts related to water resources; and implement the National Water Resources Policy. This chapter presents a discussion on the economic and financial sustainability of the National Water Resources Management System - SINGREH, as well as suggestions for its improvement.

Keywords: water resources management, SINGREH, economic and financial sustainability.

1. Introduction

The National Water Resources Management System - SINGREH provided for by Item XIX, of Article 21, of the Federal Constitution of 1988, was created by Law No. 9,433, of January 9, 1997, which in its Article 32, explains the objectives of SINGREH , namely: i) to coordinate integrated water management; ii) administratively arbitrate conflicts related to water resources; iii) implement the National Water Resources Policy; iv) plan, regulate and control the use, preservation and recovery of water resources; v) promote charging for the use of water resources.

Furthermore, in its Article 33, as amended by Law No. 9.984 / 2000, it lists the participating SINGREH bodies, namely: i) the National Water Resources Council; ii) the National Water Agency; iii) the Water Resources Councils of the States and the Federal District; iv) the River Basin Committees; v) the federal, state, Federal District and municipal government agencies whose competences are related to the management of water resources; vi) Water Agencies.

2. State of the Art of SINGREH Resource Sources

It is possible to talk about economic sustainability, in an elementary way, from the perspective of cost-benefit analysis, when the benefits are greater than the costs involved. From a financial perspective, the flows and guarantee of resources for the financing of SINGREH are related.

As sources of funds can be cited: **primary resources** (concrete sources or firm revenues, such as charging for the use of water resources, financial compensation for the result of the use of water resources for the generation of electricity and royalties; and sources potential, such as the apportionment of works of multiple use of common or collective interest and compensation to municipalities); **derived resources** (sources and financing mechanisms resulting from sectoral policies and regional development, which have interfaces with the management of water resources, as well as penalties and fines); **traditional resources** (funding resources from development banks, traditional public banks); **private resources** (private banks, Bovespa, capital markets, insurance markets, futures markets, etc.); and **international resources** (resources from international financing agencies, such as Bird, IDB, etc.).

Preliminary studies indicate that the potential for collecting charges for the use of water resources in Brazil was R$ 520 million per year, with a large concentration in the Paraná River Basin. However, the implementation of the collection entails several previous steps, in addition to relatively high transaction costs. Furthermore, there is a huge disparity in the collection potential between the different Brazilian basins. Simple collection mechanisms can, on the one hand, facilitate its implementation, but on the other hand, it can lead to a reduction in its economic efficiency, in the large number of rides or free riders (those who benefit from the policy without working in return), in possibility of catching the basin committee. Another precaution that must be taken is with the use of the outdated or inexpressive public unit price, which can be used, as well as with a high number of subsidies or exemptions to certain sectors (irrigation and Small Hydroelectric Plants - PCHs).

In the federal legal system there are several rules that deal with financial compensation due to States, the Federal District and Municipalities, due to the result, among others, of the exploitation of water resources for the purpose of generating electricity. Among them, the following stand out: Federal Law No. 7,990 / 1989; Federal Law No. 8,001 / 1990; Federal Law No. 9,993 / 2000; Federal Law No. 9,648 / 1998 and Federal Decree No. 3,874 / 2001 (with changes introduced by Federal Law No. 9,433 / 1997 and Federal Law No. 9,984 / 2000). It is true to affirm that the financial compensation is the amount that the holders of concession or authorization of plants pay to the States, the

Federal District and the Municipalities for the exploitation of hydraulic potential located in their territory.

The Federal Constitution of 1988 provided for financial compensation, as can be seen from the content of § 1 of art. 20, namely:

"Art. 20. The Union's assets are:

(...)

§ 1º. Under the terms of the law, States, the Federal District and Municipalities, as well as organs of the direct administration of the Union, are guaranteed participation in the results of the exploitation of oil or natural gas, of water resources for the purposes of generating electricity and other natural resources in the respective territory, continental shelf, territorial sea or exclusive economic zone, or financial compensation for that exploration. "

Thus, one of the forms of costing the managing body is through the amounts received by the State as financial compensation. The financial compensation referred to in Federal Law No. 7,990 / 1989 was set at an amount corresponding to 6.75% of the amount of energy produced (as specified in Article 17 of Federal Law No. 9,648 / 1998, as amended by Article 28 of Federal Law No. 9.984 / 2000), to be paid by the holder of the right to exploit the hydraulic potential (which may be the concessionaire of electricity generation services): i) to the States, the Federal District and the Municipalities in whose territories they are located installations for the production of electric energy or that have areas invaded by water from the respective reservoirs; and ii) bodies of the Federal Government's Direct Administration.

6.75% will be distributed as follows: i) 6.00% (six percent) of the value of the energy produced will be distributed among the States, Municipalities and administration bodies in the implementation of the National Water Resources Policy and the National Water System. Water Resources Management, pursuant to art. 22 of Law no. 9,433, of January 8, 1997.

In addition, the 6.00% mentioned in item (i) above must be distributed as follows: i) 45% (forty-five percent) to the States; ii) 45% (forty-five percent) to Municipalities; iii) 3% (three percent) to the Ministry of the Environment; iv) 3% (three percent) to the Ministry of Mines and Energy; v) 4% (four percent) to the FNDCT.

An interesting aspect to be addressed is that 0.75% of the percentage related to financial compensation (6.75%) will be allocated to the Ministry of the

Environment, for application in the implementation of the National Water Resources Policy and the National Water Resources Management System, under the terms of article 22, of Federal Law no. 9,433 / 1997.

Although the evolution of the legal framework at the federal level is undeniable, it is possible to detect delays and deficiencies in the process of implementing the federal laws mentioned above. Part of this difficulty comes from the text of Federal Law no. 9.433 / 1997, which refers to the regulation of a series of aspects such as, for example, uses that are not granted, which, until now, have not been regulated by Decree, approved by the Executive Branch. In this sense, it is believed that, for a better application of water resources management instruments, it is necessary to issue regulatory decrees that define peculiarities not provided for in federal laws.

Due to the lack of regulation of Federal Law no. 9.433 / 1997 and the need to define criteria not provided for in federal laws, the National Water Resources Council resolved, through Resolution, on several and relevant topics such as, for example, the performance of Basin Committees, the management of groundwater and many others, which is important for the performance of the agencies, but does not meet the need for regulation of the Federal Law through decrees, approved by the Executive Branch. Thus, it is believed that the federal rules, currently existing, will be more effective with the consolidation of SINGREH, a circumstance in which all the bodies that compose it will begin to exercise their duties to the full, in accordance to the legal rules.

Currently, a myriad of problems are observed, which, once flattened, would allow an expansion of resources in order to obtain an economic and financial sustainability of SINGREH, among which can be mentioned: i) a low amount of compensation resources financial of the States, which in 2009 reached values of the order of R $ 742 million, for SINGREH; ii) the resources of the municipalities are not linked to SINGREH (in 2009, they also reached values around R $ 742 million); iii) high contingency of resources at the federal level; in 2009, ANA (69%) and SRHU (59%); iv) the mechanisms for investing resources - ANA resources x river basin resources; v) non-payment by the Small Hydroelectric Plants - PCHs and, as a rule, subsidies and exemptions to the irrigation sector.

3. Possibility of Expanding Resources and Financial Sustainability

In addition to the recommendations mentioned above, the following possibilities for expanding the resources and financial sustainability of SINGREH also follow:

i) Implementation of the Management Contract, in accordance with Article 10, of Annex I, of Decree no. 3,962, of December 19, 2000, which proclaims that "the administration of ANA will be governed by a management contract, negotiated between its Chief Executive Officer and the Minister of State for the Environment, within the maximum period of twenty-two days following the appointment of ANA's Chief Executive Officer". The execution of the Management Contract would allow the Regulatory Agency to acquire the qualification of Executive Agency, bringing in fact, independence and administrative, financial and legal autonomy;

ii) Progress in the implementation of the collection, both in terms of the methodology used, and in the implementation of new basins;

iii) Implementation of criteria of vulnerability and resilience, instead of just reliability, in the implementation of the grant and collection;

iv) Improvement in the articulation and implementation of the National Policy on Climate Change - PNMC, especially in actions related to mitigation and adaptation to climate change and REDD + (derived resources);

v) Improvement of the articulation of water resources management with forest management (derived resources);

vi) Improvement of the articulation of water resource management with land use and conservation management (derived resources);

vii) Improving the articulation of water resources management with estuarine management (derived resources);

viii) Regulation of the National Water Resources Plan, as explained in Art. 52, of Law no. 9433/1997, which would result in a redefinition of the use of hydraulic potentials for purposes of electrical generation, no longer subordinating it to the discipline of specific sectoral legislation;

ix) Elaboration and monitoring of regulations of the Presidency of the Republic Decree, which would make it possible to define criteria for preventive rationing and the actions provided for in paragraph 2 of item X, Article 4, of Law 9.984, of July 17, 2000. This it would enable an improvement in the planning and promotion of actions aimed at preventing or minimizing the effects of droughts and floods, in articulation with the central body of the National Civil Defense System, in support of States and Municipalities, as determined by the aforementioned item;

x) Improvement in the articulation and implementation of the National Policy for Sustainable Production and Consumption - PNCS (derived and traditional resources);

xi) Improvement in the articulation and implementation of the National Solid Waste Policy (derived and traditional resources);

xii) Improvement in the articulation of the implementation of the National Sanitation Policy (derived and traditional resources);

xiii) Implementation of the Proration of Works for Common and Collective Use; Better application of the Allocation Mechanisms and Water Pacts;

xiv) Greater coordination with related MDS and MDA programs, such as the MDS / CONSEA Water Access Program (derived resources);

xv) Improvement in the implementation of Flood and Drought Alert Systems;

xvi) Greater articulation with the Navigation sector (derived resources);

xvii) Greater articulation with the Energy sector (derived resources);

xviii) The use in studies and projects of flow generation models for intermittent rivers, which reproduce the characteristics of periods of droughts and floods, which results in the bad dimensioning of the hydraulic works and in the estimation of the regularization of the reservoirs, and, consequently, in the grant and collection;

xix) Greater link with States, Municipalities, state and sector funds (FNE, etc.) and banks (NORTHEAST, BNDES, CEF, etc.).

4. Final considerations

In brief and concise terms, these are the main suggestions in order to identify sources of revenue for financing actions aimed at the integrated management of water resources, with a view to the economic and financial sustainability of SINGREH.

CHAPTER 18

Application of Multicriteria Analysis in the Study of Operating Alternatives for the Bocaina Reservoir (State of Piauí) and Development of the Region Downstream of the Reservoir (Guaribas River Basin)

Marcos Airton de Sousa Freitas – National Water Agency - ANA,
masfreitas@ana.gov.br

Abstract: The Bocaina dam, in the semi-arid region of the State of Piauí, with a capacity of 106 hm^3 of water, was built with the objective of regularizing the Guaribas River, among other uses. Over the past three decades, use for irrigation and water supply in neighboring cities has increased significantly. The objective of this study is to analyze, by means of several multicriteria methods, the alternatives of operation of the afore mentioned reservoir in order to meet multiple uses. Five alternatives for reservoir operation, socioeconomic and environmental development were analyzed. For this, four different multicriteria techniques were used: i) weighted average method (WAM); ii) compromise programming (CP); iii) Promethee method with weighted averages (Promethee_WAM) and iv) Promethee method.

Keywords: multicriteria methods, decision-making processes, reservoir operation.

1. Introduction

The decision-making process in the area of water resources currently involves multiple objectives and multiple decision-makers (river basin committees, consortia, etc.). In general, conflicts of interest are established between groups with different views about the goals to be adopted in the planning and management of water resources. For groups linked to pure and simple economic development, the net economic benefit should be maximized, since economic values express the interest of society. On the other hand, environmental groups preach the preservation of the environment in its natural form and are opposed to any intervention that may transform it. Between these two extreme possibilities there is a set of compromise solutions (Braga & Gobetti, 1997).

According to Cohon & Marks (1975), multi-objective analysis techniques can be divided as follows:

Techniques that generate the set of non-dominated solutions:

- Weighting method (Zadeh, 1963);
- Method of restrictions (Zadeh, 1963);
- Linear multi-objective method (Philip, 1972).

Techniques that use an advance articulation of preferences:

- Multidimensional utility function method (Keeney and Raiffa, 1976);
- Goal programming method (Charnes and Cooper, 1961);
- Electre Method (Roy, 1971);
- Promethee method (Brans and Vincke, 1985);
- Surrogate trade-off method (Haimes and Hall, 1974);
- Analytic Hierarchy Process (Saaty, 1977);
- Q-Analysis technique (Hiessl et al., 1985).

Techniques that use a progressive articulation of preferences:

- Step method (Benayoun et al., 1971);
- Compromise Programming method (Yu, 1973)

The Bocaina Reservoir (Figure 01), with a capacity of 106 hm^3 of water, was built under the responsibility of DNOCS - National Department of Works Against Drought - and executed by the Engineering and Construction Battalion, with the objective, among other aspects, of regularize the Guaribas River.

The area estimated by the viability studies of the Bocaina Reservoir reached 2,000 (two thousand) irrigable hectares, according to the Bocaina Dam - Hidroterra SA Executive Project - located in the municipalities of Bocaina, Sussuapara and Picos, with excellent physical and chemical conditions for intensive irrigation use , at a relatively low cost, considering that there was no cost for deforestation, systematization and settlement of producers.

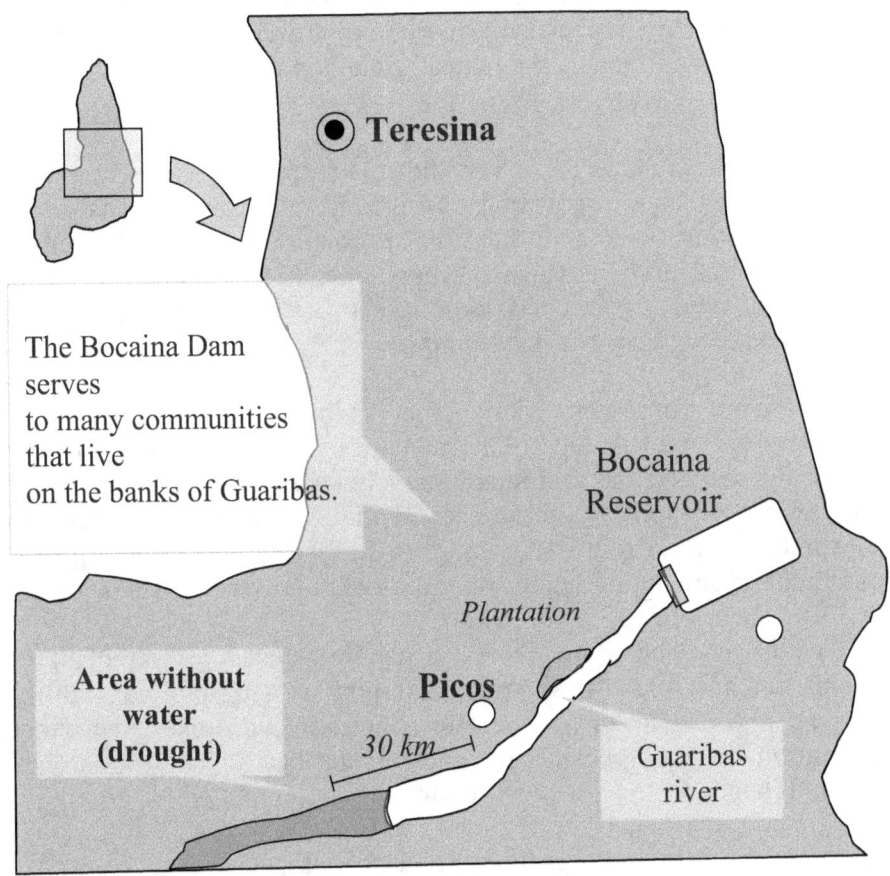

Figure 1: Bocaina Dam location.

The study of alternatives (A1 to A5) was carried out in two stages:

  • Use of rainfall-runoff hydrological models (Freitas, 1991), stochastic streamflow generation models (Freitas, 1995 and 1997) and Monte Carlo analysis, as well as models for simulation / optimization of the reservoir operation;
  • Use of multicriteria techniques, such as weighted averages method, compromise programming (CP) method and Promethee method.

2. Simulation of the Bocaina Reservoir Operation

The current complexity of problems related to the management of water resources requires the use of instruments and techniques capable of assisting in decision making, especially in periods of scarcity. Among these techniques, two deserve mention: mathematical simulation and flow network modeling (Azevedo et al., 1997).

Mathematical simulation is a very flexible technique and is therefore widely used, however, as a disadvantage of not offering users the chance to restrict decision-making space and, therefore, problem solving is achieved through a trial and error process. On the other hand, the network flow modeling, in addition to reducing the decision space to a set of viable solutions, can be optimized through linear programming.

One of the most used network flow models in the world is the MODSIM model, developed by John Labadie, at Colorado State University, in the United States. In this study, two versions were used, namely MODSIMP32 and ModSimLS, which incorporate a graphical interface, in Windows environment, developed at the University of São Paulo - USP, by Prof. Dr. Rubem La Laina Porto. Next, the methodology for simulating the operation of the reservoir will be presented.

The water availability in a river is generally used to water supply to the population and industries; animal supply; irrigation of agricultural areas; ecosystem conservation; wastewater dilution; hydro-energy production and navigation. In the specific case of the Guaribas River, downstream of the Bocaina Dam, only the last two uses are not contemplated.

The simulation method uses the continuity equation:

$$S_t = S_{t-1} + \overline{Q}_t \Delta t + (P_t - E_t).K.\overline{A}_t + Qs_t \Delta t$$

(01)

with

S_t and S_{t-1} are, respectively, storage at times t and t-1;
$\overline{Q}_t$ is the average flow into the reservoir between the mentioned intervals;
P_t is the precipitation in the period;
E_t is the potential evaporation on liquid surfaces in the period;
$\overline{A}_t$ s the average area of the reservoir in the period;
K is the unit conversion factor;
Qs_t is the average outlet flow;
Δt is the time interval.

St storage must be between the limits Si and Sm. The first represents the minimum volume of the reservoir (dead volume), from which demand cannot be met, the second being the maximum capacity of the reservoir. In the simulation of the reservoir operation, the process is initiated considering the average accumulated volume of the dam. Then, several simulations are made to define regularized volumes, associated with various guarantees or risks.

In relation to irrigation, simulations were made for several demand areas, based on the estimated area potential for the downstream valley, forecasting monthly demands based on percentages of annual demand, proportional to the monthly deficits determined based on the water balance in the area.

The failure risk was calculated based on the number of failures in the period of interest divided by the number of time intervals for that period, that is, the simulated period. Failure is adopted when demand is not met. The failure probability is therefore estimated by:

$$Failure = \frac{NF}{NIT}$$

where   NF is the number of time intervals with failures;
     NIT is the number of total time intervals for the demand period.

The simulation model of the reservoir operation used was the one shown in Figure 02, developed by Prof. Dr. Fontane, at the University of Colorado. The model uses Excel's Solver to maximize the annual demand met for the various uses in the basin (irrigation, human supply and animal feed) and to determine the useful (active) volume needed to meet a given demand (reservoir project). The minimum and maximum volume of the Bocaina Dam is 10 and 106 hm^3, respectively. The maximum annual demand met for the simulated series was 22.27 hm^3.

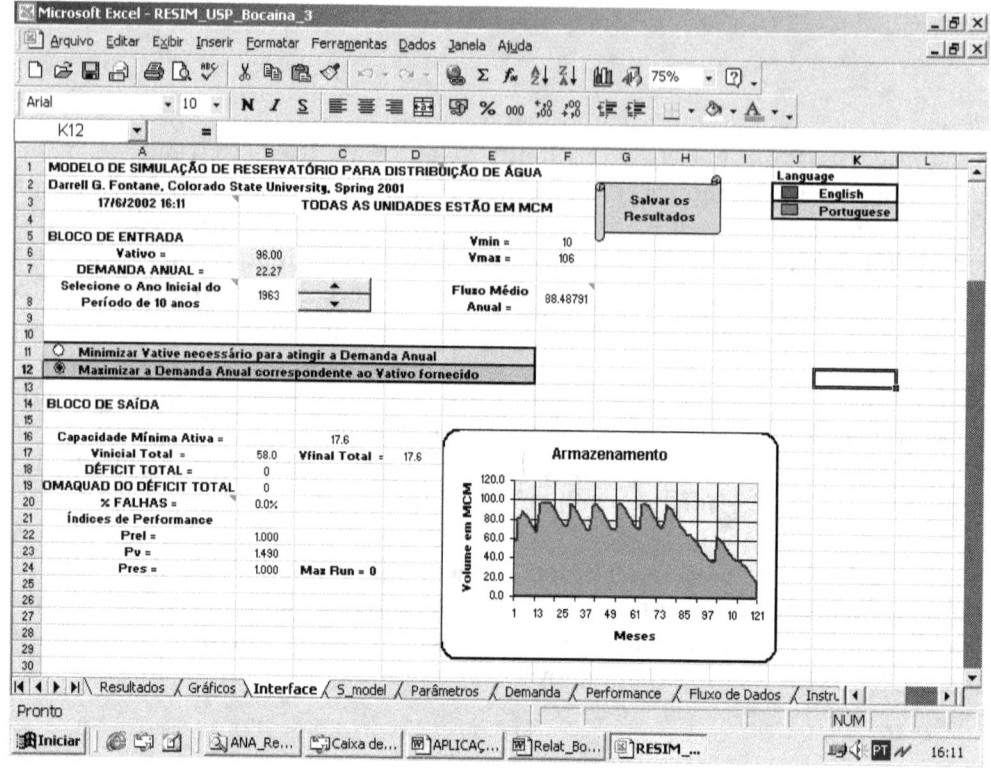

Figure 2: Reservoir operation simulation model.

3. Multicriteria Analysis in the Alternatives Study

For the second stage, the MCDA spreadsheet - Multi-Criteria Decision Analyzes (Figure 03), also developed by Prof. Dr. Fontane, from the University of Colorado. Five alternatives (A1 to A5) were considered for dam operation / economic and environmental development in the region. The alternatives were evaluated according to five criteria: i) irrigation; ii) human and animal supply; iii) the environment; iv) safety and reliability of the dam's operation and v) economic development. Up to five sub-criteria were established for each of the above criteria. Figure 04 shows the values of the criteria and sub-criteria for each alternative. Five alternatives of relative importance between the criteria were analyzed (G1 to G5). The relative weights between the criteria and sub-criteria were raised based on surveys with specialists in the area and interviews between users.

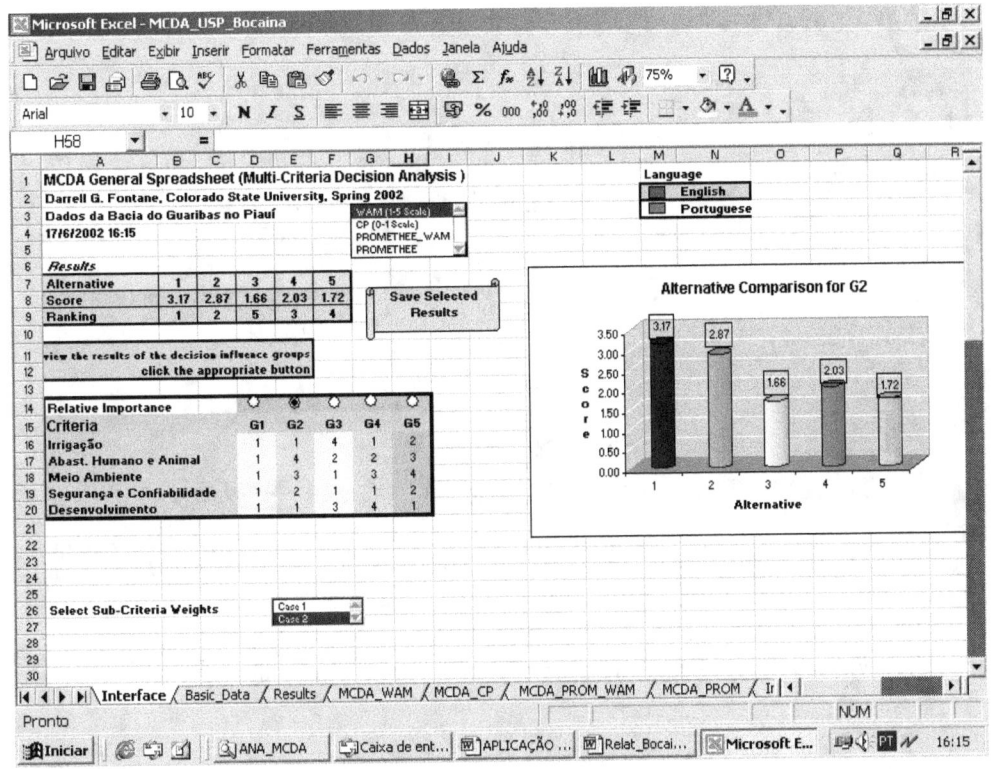

Figure 3: Decision making multicriteria models.

4. Conclusões e Recomendações

Four different techniques were used: i) weighted averages method (WAM); ii) compromise programming (CP); iii) Promethee method with weighted averages (Promethee_WAM) and iv) Promethee method. As a result, there are the "scores" and the ranking (hierarchy) among the various alternatives. In the Bocaina Basin example, the following priority was found: A1 → A2 → A5 → A3 → A4.

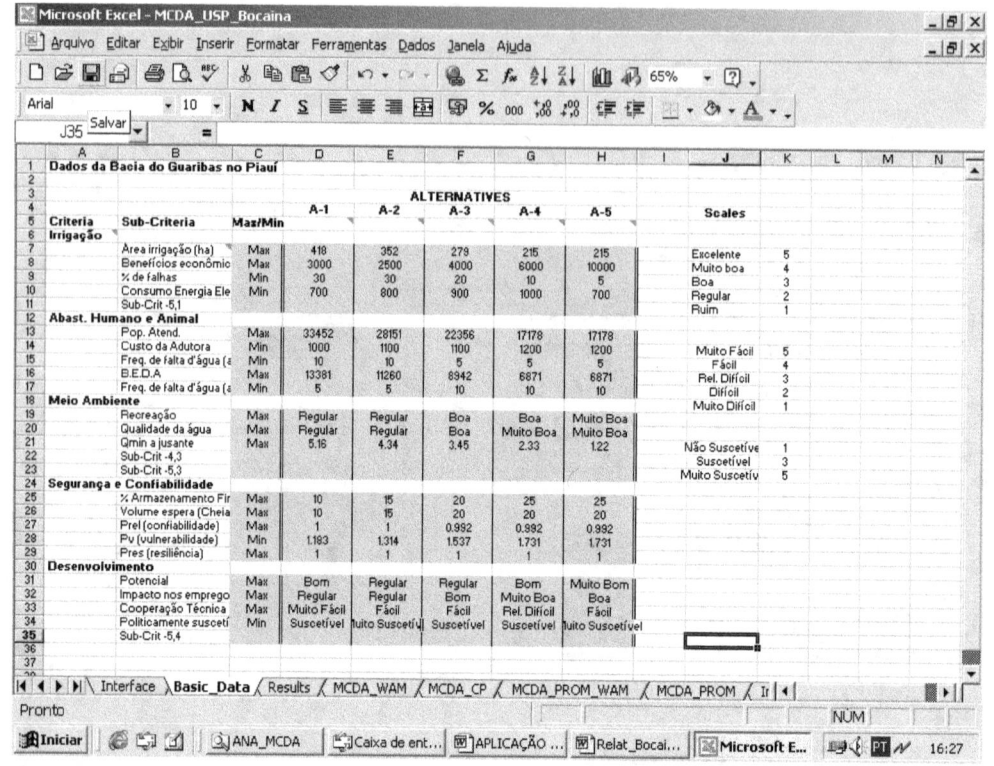

Figure 4: Values of criteria and sub-criteria for each alternative.

5. References

AZEVEDO, L.G.T.; PORTO, R.L.L.; ZADEH FILHO, K. – Modelos de Simulação e de Rede de Fluxo, in: Técnicas quantitativas para o gerenciamento de recursos hídricos, R. La Laina Porto (org.), Editora da Universidade UFRGS/ABRH, 1997.

BENAYOUN, R; MONTGOLFIER, J.; TERGNY, J.; LARITCHEV, O. – Linear programming with multiple objective functions: step method, Mathematical Programming, vol. 1, n.3, 1971.

BRAGA, B.; GOBETTI, L.- Análise Multiobjetivo, in: Técnicas quantitativas para o gerenciamento de recursos hídricos, R. La Laina Porto (org.), Editora da Universidade UFRGS/ABRH, 1997.

BRANS, J. P.; VINCKE, P. – A preference ranking organization method, Management Science, vol.31, n.6, 1985.

CHARNES, A., COOPER, W. – Management models and industrial applications of linear programming, New York, John Wiley, 1961.

COHON, J.L.; MARKS, D. H. – A review and evaluation of multiobjetive programming techniques, Water Resources Research, v.11, n.2, 1975.

FREITAS, M.A.S. - Considerações Sobre Modelos Determinísticos Chuva-Vazão Aplicados às Bacias do Semi-Árido Brasileiro, Dissertação de Mestrado, Universidade Federal do Ceará - UFC, Fortaleza, 1991.

Freitas, M.A.S. - SAGE (Stochastische AbflussGEnerierungsmodelle), Handbuch, Institut für Wasserwirtschaft, Hydrologie und landwirtschaftlichen Wasserbau, Universität Hannover, Alemanha, 1995.

Freitas, M.A.S. - Regionale Dürreanalyse anhand statistischer Methoden und Neuro-Fuzzy-Systemen mit Anwendung für Nordost-Brasilien, Doctoral Dissertation, University Hannover, Germany, 1997.

HAIMES, Y. Y.; HALL, W.A. – Multiobjectives in water resource system analysis: the surrogate worth trade off method, Water Resources Research, vol. 10, n.4, 1974.

HIESSL, H.; DUCKSTEIN, L.; PLATE, E.J. – Multiobjective Q-Analysis with concordance and discordance concepts, Applied Mathematics and Computation, vol.17, 1985.

KEENEY, R.; RAIFFA, H. – Decisions with multiple objectives: preferences and value trade-offs, John Wiley, New York, 1976.

PHILIP, J. – Algorithms for the vector maximization problem, Mathematical Programming, vol.2, 1972.

ROY, B. – Problems and methods with multiple objective functions, Mathematical Programming, vol.1, 1971.

SAATY, T.L. – A scaling method for priorities in hierarchical structures, Journal of Mathematical Psicology, vol.15, 1977.

YU, P. – A class of decision problems for group decision problems, Management Science, n.19, 1973.

ZADEH, L. – Optmality and no scalar valued performance criteria, IEEE Transactions on Automatic Control, v.59, n.8, 1963.

ABOUT THE EDITOR

MARCOS AIRTON DE SOUSA FREITAS

Post-graduation in Civil Engineering - Dipl. Ing. - Universität Hannover. M. Sc. Civil Engineering (UFC). MBA in Public Management (ENAP). Graduation in Civil Engineering (UFPI). Specialist in Water Resources at the National Water Agency - ANA, since 2001. University Professor, from 1990 to 2016. Civil Engineer / Consultant (1985-2000). Coordinator and professor of several Postgraduate courses (Software Engineering; Water Resources Management; Environmental Management). More than 100 technical-scientific publications in journals and annals of national and international symposia and more than 50 books and 30 book chapters, technical and literature, in authorship and co-authorship, in 6 languages. Participated in the preparation of the National Water Resources Plan - PNRH, the São Francisco River Basin Plan - PDRHSF, GeoBrasil Water Resources (UNDP), GEF São Francisco, among others. Founder and Former Technical and Scientific Director of the Association of Servers of the National Water Agency - ASÁGUAS. Founder and Former Adviser of the National Association of Regulatory Specialists - ANER. Member of the Brazilian Association of Water Resources - ABRH. Founder of the Astronomy Association of Piauí - APA (1982). Affiliated to the National Writers Association – ANE, to the National Writers Union – UBE and to Academia de Letras do Brasil - ALB.

Currículo Lattes: http://lattes.cnpq.br/8217649125563058

Page on Amazon: https://www.amazon.com/Marcos-Freitas/e/B0735V21TV

ABOUT THE AUTHORS

GABRIEL BELMINO FREITAS

Graduation in Economics at the University of Brasília (UnB), with final work addressing the inflation forecast using machine learning methods. He interned at Instituto Banco Palmas and Ábaco Contabilidade, in Fortaleza, respectively, in 2015 and 2016. He served as a volunteer at AIESEC, in Romania, in 2018, teaching children and adolescents in English about Sustainable Development Goals - SDGs). From October 2019 to February 2020, he has been an economic development intern at ECLAC / UN, Brasília office. From May 2020 to September 2020 was an assistant researcher on the CoronaNet Research Group (Munich). And since October 2020, Master of Science student in Applied Economics at Leopold-Franzens Universität Innsbruck. Co-author of the book "Inteligência Artificial e Machine Learning: teoria e aplicações (ISBN: 979-86-2141-717-8).

https://www.linkedin.com/in/gabriel-belmino-freitas-90392119b/

SANDRA REGINA AFONSO

She has a degree in Agronomic Engineering from the University of São Paulo (1996), a master's degree in Forestry Sciences from the University of Brasília (2008) and a doctorate in Forestry Sciences from the University of Brasília (2012). Since 2010, technologist researcher at the Brazilian Forest Service. She has experience in the forestry area, having worked mainly on the following themes: forest bioeconomy; management and conservation of the *cerrado* (Brazilian savanna); non-timber forest production; environmental education; forest policies; and development of forest productive chains. Author of the book "Políticas Públicas de Incentivo à Produção Florestal Não Madeireira no Brasil / Public Policies to Encourage Non-Timber Forest Production in Brazil" (Amazon, 2017; ISBN-13: 978-1521996188) and co-author of the book "Bioeconomia da Floresta: Conjuntura da Produção Florestal Não Madeireira no Brasil / Forest Bioeconomy: Situation of Non-Timber Forest Production in Brazil (MAPA/SFB, ISBN: 978-85-7991-132-3). Since 2020, also associated researcher at the University of Brasília – UnB.

Currículo Lattes: http://lattes.cnpq.br/7967812965766700

https://www.escavador.com/sobre/3272974/sandra-regina-afonso

PAULO BRENO DE MORAES SILVEIRA

Graduation in Chemical Engineering (University of São Paulo - USP), in 1974. M.Sc. Chemical Engineering (University of São Paulo - USP), in 1982. He worked as a researcher at the Institute of Technological Research (Instituto de Pesquisa Tecnológica) - IPT, in São Paulo, from 1975 to 1995. Specialist in Water Resources at the National Water Agency - ANA, from 2001 to 2019. Former Professor of Hydraulics and Sanitation in the Department of Civil Engineering at Maringá State University, in the State of Paraná, from 1998 to 2001.

MARCIO ANTONIO SOUSA DA ROCHA FREITAS

He holds a BA in Law from the Federal University of Piauí - UFPI (1996), a BA in Agronomic Engineering from UFPI (1988), a Specialization in Constitutional Law (1999) and a Specialization in Water Resources and Environment Management (2002) from UFPI. Master in Agricultural Engineering from the Federal University of Ceará - UFC (1993) and PhD in Ecology and Natural Resources from the UFC (2014). He is currently an Adjunct Professor at the State University of Piauí-UESPI and the Faculty of Higher Education of Floriano - FAESF. Former President of the Environmental Commission of the Brazilian Bar Association - Piauí Section - OAB-PI, former Coordinator of the Environment Committee of the Regional Council of Engineering and Agronomy - CREA-PI, former Member of the State Council of Water Resources of the State of Piauí - CONSERH, former member of the Municipal Council for the Environment and Water Resources of Teresina - COMDEMA, former Director of the Piauí Engineers Union - SENGE, Regional Representative of the Brazilian Water Resources Association - ABRH and Brazilian Association of Sanitary and Environmental Engineering - ABES, current President of the Velho Monge Foundation - Rio Parnaíba Vivo. Works on the following themes: hydrosedimentology, ecosystem ecology, watershed, management of the environment and water resources, environmental sanitation, environmental law, agrarian law and urban law.

Currículo Lattes: http://lattes.cnpq.br/0505387798204439

Sumário

ABOUT THE AUTHORS